THE MAKING OF
IRELAND'S
LANDSCAPE

SINCE THE ICE AGE

VALERIE HALL

The Collins Press

First Published in 2011 By
The Collins Press
West Link Park
Doughcloyne
Wilton
Cork

© Valerie Hall 2011

Valerie Hall has asserted her moral right to be identified as the author of this work.

British Library Cataloguing in Publication Data
Hall, Valerie.
The making of Ireland's landscape : since the Ice Age.
1. Geomorphology--Ireland. 2. Landscapes--Ireland.
3. Nature--Effect of human beings on--Ireland.
I. Title
554.1'5-dc22
ISBN-13: 9781848891159

Cover image is © Eamonn O'Donoghue
Design and typesetting by Burns Design
Typeset in Adobe Garamond
Printed in Spain by GraphyCems

Cover photographs
Front: Aerial view of the Hill of Tara, County Meath (Eamonn O'Donoghue/ World Monuments Fund: Hill of Tara proposal, photograph EOD, 2006); *Back*: One of many Neolithic field walls at Céide exposed when the overlying blanket peat was removed (Jonathan Pilcher).

Photographs on p. i: beech leaves (courtesy of stock.xchng); *pp. ii–iii*: Burren landscape (courtesy of stock.xchng).

THE MAKING OF
IRELAND'S
LANDSCAPE

SINCE THE ICE AGE

VALERIE HALL has had a life-long interest in plants and the Irish landscape.
A botanist before specialising in the history of the Irish environment, she contributed
to radio and television programmes. After a lifetime of field observation
and studies at Queen's University, Belfast, she is now Professor Emerita
of Palaeoecology. She was joint author of *Flora Hibernica* (2001).
Email: v.hall@qub.ac.uk

CONTENTS

IN MEMORY OF GEORGE
AND FOR MY DEAR GRANDCHILDREN
THAT THEY MAY LEARN TO LOVE THEIR NATIVE LAND

PREFACE AND ACKNOWLEDGEMENTS

This book is for people interested in the Irish landscape and how it has developed since the last Ice Age ended. Over the years I have been privileged to speak to historical and natural historical societies throughout Ireland on topics linked to past environmental change. At the end of each lecture there was a time set aside for questions, during which I was often asked what books I would recommend so that members could read more about the subject. My answer never varied: *The Way That I Went*, published in 1937, the magisterial work of the botanist Robert Lloyd Praeger, and Frank Mitchell's *The Irish Landscape*, the first edition of which was published in 1976. Much that is new about Ireland's past environment has been discovered since those influential books were written but the findings are published mostly in scholarly papers or reports. It seemed to me that it was time for a new book to be written for a non-specialist readership, a book which would combine the classic findings with the new discoveries.

That I should tackle writing the book made me smile, in anticipation of intellectual delights. The business of drafting the chapters and planning the illustrations made me smile less. I had a great deal of reading to do but even that hefty task became a pleasure as I became reacquainted with the classic papers and met, for the first time, many publications that I may not have encountered otherwise. To illustrate the text I have selected many photographs from my own collection that show common native plants and modern ecosystems, in which the reader may find reflections of past environments. The text of the book draws overwhelmingly on published findings, by scholars working in the disciplines of botany, geography, archaeology, climatology, geology, zoology and various historical disciplines, since the nineteenth century. Some of the most exciting information had come from the pens of 'amateurs' publishing in, for example, the *Irish Naturalists' Journal* and who were as keen on 'finding out more' as are the members of the local historical societies and field clubs today.

From the middle decades of the twentieth century the Danish scientist Knud Jessen and Ireland's leading landscape expert, Frank Mitchell, led the way in gathering and interpreting evidence for Ireland's environmental past. Other scientists soon followed, namely Mike Morrison, Nick Stephens and Alan Smith, and enlightened our understanding of ancient woodlands and the emergence of agriculture, mostly through ground-breaking pollen analytical investigations. Bill Watts and then Gurdip Singh tackled some of the first studies of vegetation change in Ireland as the last Ice Age ended. More recently we are indebted to the outstanding scholarship of Eric Colhoun, Michael O'Connell, Peter Coxon, Tom Hayden, Gerry Doyle, Richard Bradshaw, Stephen Waldron, Fraser Mitchell, Karen Molloy, John Feehan and their colleagues and students who have studied animals and plants that inhabited Ireland in the past. With equal excellence Peter Woodman, Robert Devoy, Julian Orford, Peter Wilson and their co-workers have welded together findings from archaeological and palaeoenvironmental investigation, while John Sweeney and David Jewson have made significant contributions to our understanding of early climate change and the biology of ancient lakes. Without the knowledgeable staff and the collections in the

National Museum of Ireland, the Museums and Galleries of Northern Ireland, the Royal Irish Academy and the Society of Antiquaries of Ireland, the work of academics would have been hampered greatly.

Closer to home, at Queen's University Belfast, I have been helped in my task by pioneering workers, in particular Adelaide and Ian Goddard. I have drawn much on studies performed by my colleagues of many years especially Jonathan Pilcher and Mike Baillie who, along with Paula Reimer, Gerry McCormac, Keith Bennett, Gill Plunkett, Nicki Whitehouse, Finbar McCormack, Emily Murray, Meriel McClatchie, Lisa Coyle, David Brown and David Weir, have pressed on with their research into tree-ring dating, ancient pollen and insect studies, radiocarbon dating, seed, plant fossil and animal bone analyses from archaeological sites and, most recently, pioneering studies based on microscopic Icelandic volcanic ash trapped in the Irish peat and lake sediments.

It is not only people working in Ireland who have generated findings about Ireland's ancient environment. In the United Kingdom, the Netherlands and Australia, Rick Battarbee, Russell Coope, Mike Walker, John Lowe, Kevin Edwards, Keith Barber, Chris Caseldine, Frank Chambers, Geoffrey Blackford, Bas van Geel and John Dodson have generated much that is scholarly and pertinent. Their exemplary work has stimulated scientists elsewhere to further meticulous research on global past climate and environmental change, thus reinforcing Ireland's place on the world map of scientific excellence, at this time when efforts to assess rates of future climate and landscape change has become crucial. The errors in the text are mine alone.

There are the other friends and family who have offered me help and support as the book progressed. Siobhan Geraghty and Robin Govier, my friends for 'donkey's years', provided thoughtful comments as well as meals and cups of coffee. I knew that Vincent and Peggy Geraghty would take great interest in all the stages of writing and I thank them for their unflagging support. George Cunningham, Dom Laurence Walsh and the Cistercian monks of Mount St Joseph Abbey, along with the 'regulars' at the Roscrea conferences, asked the really difficult questions. During the academic year 2009–10, when I was the Parnell Fellow in Magdalene College, University of Cambridge, the Master and Fellows gave me their friendship and support as I wrote the final chapters of the book. Closest to my heart, until his death in late 2007, my beloved husband George encouraged me to keep the whole project going. Throughout it all my dear mother Evelyn and sister Arnette, daughters Fiona and Roisin and sons-in-law Eamonn and Jamie have helped me unstintingly. Grandaughter Cora and I have enjoyed many lively discussions about mammoths, not that they have been in Ireland all that recently. I thank all of you.

VALERIE HALL, February 2011

LIST OF ILLUSTRATIONS

Selected areas and sites of past environmental significance

Legend:

— County boundary

Site Locations
- o ○ Rural/Urban
- △ Mountain/cliff

B'MBS MORE — BALLYMACOMBS MORE
R. — THE RODDANS
ARDS PEN. — ARDS PENINSULA

N

Scale:
0 10 20 30 40 50 miles
0 20 40 60 80 km

Map labels:

MULROY BAY, LOUGH SWILLY, GIANT'S CAUSEWAY, WHITEPARK BAY, DUNFANAGHY, LOUGH FOYLE, COLERAINE, GARRY BOG, LONDONDERRY, ANTRIM, DONEGAL, LIFFORD, FALLAHOGY BOG, PLATEAU, SPERRIN MTNS, B'MBS MORE NEWFERRY, ANTRIM, LOUGH MULLAGHLAHAN, BEAGHMORE, SLIEVE GALLION, SLUGGAN BOG, MEENADOAN, BELFAST, SLIEVE LEAGUE, TYRONE, ARDS, BALLYSHANNON, NAVAN FORT, LAGAN, DOWN, R., PEN., CARROWNAGLOGH, CEIDE FIELDS, BEN BULBEN PLAT., FERMANAGH, ARMAGH, WOODGRANGE, SLIEVE CROOB, WHITE BOG, HANGING ROCK, CUILCAGH MTN., SHANNON POT, MONAGHAN, MOURNE MTNS., STRANGFORD LOUGH, SLIGO, LEITRIM, CAVAN, LOUTH, MAPASTOWN, NEPHIN BEG MTNS, EMLAGH BOG, BOYNE, DROGHEDA, NEWGRANGE/KNOWTH, CLARE IS., MAYO, ROSCOMMON, LONGFORD, MEATH, CONNEMARA, CORLEA BOG, SCRAGH BOG, CLONYCAVAN, WESTMEATH, GALWAY, CLONMACNOISE, CROGHAN, CARBURY BOG, LIFFY, DUBLIN, CLONFERT BOG, OFFALY, KILDARE, BALLYBETAGH BOGS, LOUGH BOORA, GLASHBAWN, SHANNON, ARAN IS., BURREN, DUNAMASE, WICKLOW MTNS., LOUGH NAHANAGAN, LOUGH DERG, LAOIS, WICKLOW, CLARE, SILVERMINE MTNS., CARLOW, LIMERICK, LOUGH GUR, LITTLETON BOG, KILKENNY, TIPPERARY, WEXFORD, FERRITER'S COVE, TRALEE, CRAG CAVE, CASTLEPOOK CAVE, LIMERICK, DINGLE PEN., KERRY, KILLARNEY, BALLYORAN BOG, WATERFORD, CARNSORE POINT, VALENCIA IS., ROSS IS., MUCKROSS & REENADINNA, CORK, MAGILLICUDDY'S REEKS

CHAPTER 1

SOURCES OF EVIDENCE: CLOCKS AND FOSSILS

KEY ISSUES:

gathering the lines of evidence;

constructing time frames.

TRACING THE STORY

Told in the chapters that follow, the story of Ireland's changing landscape took about 14,000 years to unfold, beginning with an account of happenings as the last Ice Age ended, finishing in the opening decade of the twenty-first century AD and telling of the landscape's response to natural forces and the influences of people. The account is based on a wide range of evidence and relies on secure methods of dating past environmental change and we begin by looking at methods whereby we trace the passage of time.

The story of environmental change over the last two hundred years can be traced in unsurpassed detail through dated accounts and instrumental records. There are numerous accounts that describe landscape change from the seventeenth century to the present. Maps and photographs reveal how towns developed and show how farming altered the landscape. At the astronomical observatory at Armagh readings from thermometres, barometres and rain gauges preserve the records for sunshine and storm from the mid-nineteenth century until the present day.

In contrast to this super-abundance of information about conditions during recent centuries, descriptions of the landscape during the eighteenth and seventeenth centuries are less abundant, and further back in time the records are even more scarce. For earlier historic periods we must rely on deducing information from the descriptions available rather than on detailed accounts. For example, from comments in seventeenth century accounts of wood products and the numbers of woodmen engaged in timber preparation, the nature and extent of woodland may be surmised. Oak woodland gets frequent mention because it yielded valuable timber and bark goods but only occasionally do other types of woodland receive comment.

As well as woods, there are accounts of bogs in a few of the letters that have survived from this period. Bogs and woods were known to shelter the rebellious Irish, who would be better 'gotten rid of'. An unnamed author records in the Calendar of State Papers for 1601: 'the woods and bogs are a great hindrance to us and a help to the rebel . . . If the country is quieted by cutting off the principal rebels much good could be done to the bogs by our labour and by the Irish churls felling, dressing and burning the trees in heaps.' The wild lairs of rebels or 'woodkern' thus receive caustic comment, although the location of these dangerous places was rarely mentioned.

In writings before the fifteenth century landscape details become increasingly patchy, dwindling almost entirely by AD 1000. For information about earlier millennia, therefore, instead of relying on records made by people, we turn to the evidence secreted within the countryside itself. The long story of Ireland's past landscape is not in history's pages but locked away within the peat bogs and lake sediments in which Ireland abounds.

TRACING THE FOSSIL RECORD

When tracing relatively recent landscape change (for in geological terms the last 14,000 years are very recent indeed) studies of fossils are of fundamental importance but the term 'fossil' is not strictly accurate when investigating the Irish landscape since the great ice sheets melted. It is geologists who make use of 'real fossils', those parts or impressions of ancient plants or animals that have become petrified, meaning turned to stone. In contrast, studies of the biological change in the last 14,000 years draw on the evidence from the semi-decomposed parts of plants and animals that have been preserved under conditions in which fungal and bacterial decay is almost halted through the exclusion of air or water.

1:1 Fungi will decompose dead matter such as wood where water and oxygen are available

Under desert conditions, for example, dead plant or animal matter remains extremely dry and thus almost immune from attack by decomposing organisms. Conversely, sub-fossils may form where there is plenty of water that is almost devoid of oxygen. In Ireland it is this latter circumstance that has preserved vast amounts of dead plant and animal material. Bogs, fens and lake mud that have developed over thousands of years are soaked in water from which dissolved oxygen has long-departed. Fragments of dead plants or animals contained in the peat or mud remain only partially decomposed, even after thousands of years. Peat that is composed of dead *Sphagnum* mosses has an additional property because the moss cell-walls produce acids that further preserve dead plants. Thus peat and lake mud store the sub-fossils of the plants and animals that once lived in or close by the bog or lake. The peat or lake mud accumulated as the centuries passed, with the oldest material at the bottom of the deposit and the youngest at the top.

TRACING THE PLANT RECORD

The plant sub-fossils captured in peat range in size from microscopic pollen grains and fungal spores, through to fist-sized clumps of dead fibrous cotton grass (*Eriophorum* sp.) to larger sub-fossils including twigs and branches with the largest of all being the huge trunks and roots of bog timber, especially Scots pine (*Pinus sylvestris*) and oak (*Quercus* sp.).

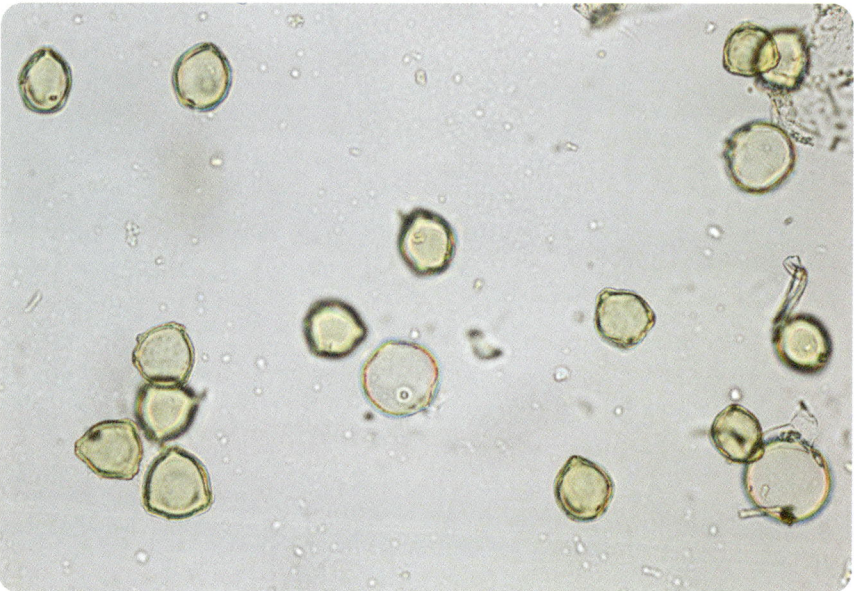

1:2 Light micrograph of pollen grains of grass and hazel. The grass pollen is thin-walled with a small pore in the side of the grain. The hazel pollen is broadly triangular in shape.

1:3 Cotton grass growing on the surface of a bog. On death, the plants' stems and leaves form a fibrous peat

1:4 Bog pine timber preserved in a peat bog

Broadly speaking, the sub-fossils that are visible with the naked eye are the 'macro-fossils', with those that can be seen only through a microscope known as the 'micro-fossils'. From the bottom of a bog to its top the pollen and woody fossils echo the changes in local vegetation through time.

The woody remains in bogs provide a useful source of information about the trees that grew at or near the site but if the wood fragments have become mushy they can be difficult to identify using only the naked eye. Small pieces of soft, partially-fossilised wood usually require laboratory identification. By examining the wood fragments' microscopic structures and then comparing these with named modern timbers, the identity of the old wood can be established.

In some places the fossilised wood reveals the tree species that once grew in ancient woodlands. Today, Clare Island in County Mayo is almost treeless but in former times trees were more abundant on the island and there was once woodland on the windswept Knockmore Cliffs. This is confirmed by birch (*Betula* sp.) branches protruding from cuttings in the thin peat that cover the cliff top. Throughout the country, at the bottom of upland blanket peat cuttings, big chunks of wood may be found, proving that there had once been woodland in places now completely devoid of trees.

The fossil timbers of yew (*Taxus baccata*), elm (*Ulmus glabra*) and holly (*Ilex aquifolium*) are less common and mostly restricted to the midland bogs. Stumps and trunks are all that remain of huge yews that grew at a time when these great wetlands had been drier. In times in the past, when Ireland's climate was less wet, bogs dried enough for yew and pine to colonise their surfaces. In a country as rain-soaked as Ireland, it is surprising to learn that fire once played a part in woodland ecology. By the sides of cut-over lowland bogs there may be found ancient pine trunks that display patches of burnt bark and charred scars deep within the wood. The new wood that surrounds the scar shows that the tree survived some ancient forest fire. Lightning may have caused these fires, burning the resinous pine wood, which will char even when damp.

A CALENDAR WITHIN THE FOSSIL PLANT RECORD

It is the bog timbers that contribute so much to the story of Ireland's changing landscape through time because their annual rings form a precise calendar. The science of time measurement based on tree-rings is called dendrochronology and the natural calendar formed by the tree-rings in Irish oak timbers is precise to the year,

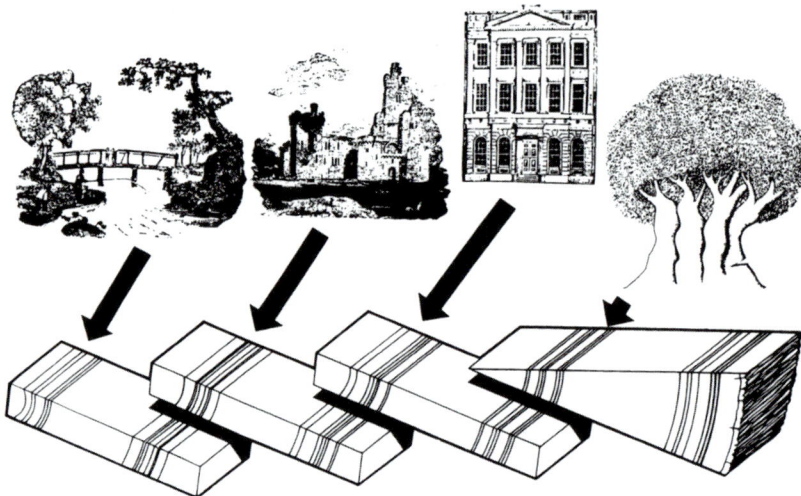

1:5 Schematic representation of the principle behind dendrochronology

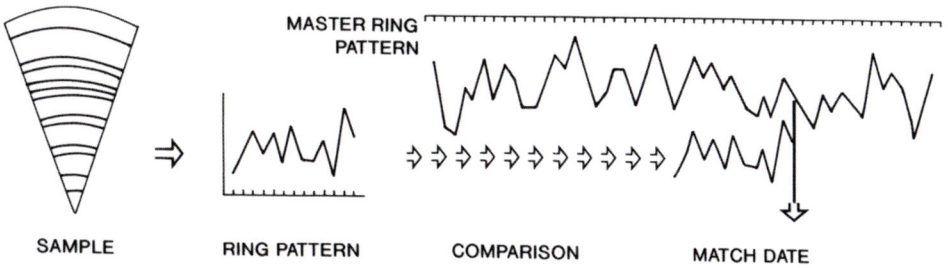

1:6 Schematic representation of the dating process

for over 7,000 years. The oak tree-ring calendar relies on the unique pattern of wide and narrow annual rings made by the tree as it grows each year in response to changes in its surroundings. To build the calendar the pattern of wide and narrow annual growth rings in living trees is overlapped with those from trees recently dead and then the oldest ring patterns in those is overlapped with the pattern of rings in even older wood.

Old oak timbers can be found in medieval buildings such as the castle at Trim or the cathedral at Kilkenny. In turn, timbers of the medieval period have their ring-width patterns overlapped with still older structures, such as those from early medieval horizontal corn mills and the artificial lake islands, known as *crannógs*. Working backwards in time, timbers from archaeological structures of the first millennium BC may then overlap with the last of the timbers from the bog oak woods that disappeared about 2,300 years ago. Working across increasingly ancient oak timbers from about 4,000 oak wood samples, annual ring-width patterns have been reconstructed for over 7,000 years, allowing us to rebuild the tree-ring pattern of an Irish oak tree, had one lived for over 7,000 years.

Tree-ring studies have contributed enormously to the dating of old, natural features as well as those made by people. For example, during an archaeological excavation at Emain Macha (now known as Navan, in County Armagh) there was discovered the stump of the central oak post of the large building that once stood on the site. The date of the post was fixed at 95 BC providing a precise felling date for the tree whose trunk was in the centre of a mysterious building at the heart of a site of international, cultural significance.

TRACING THE STORY THROUGH THE SMALLER PLANT FOSSILS

Other plant parts have a contribution to make to the story of Ireland's ancient past. Fossilised leaves of dwarf willow (*Salix herbacea*) occur in the deposits laid down as the Ice Age ended. In Belfast, on the banks of the River Lagan at Annadale, the dwarf willow's fossil leaves have been found in the silty rubbish dumped by a former

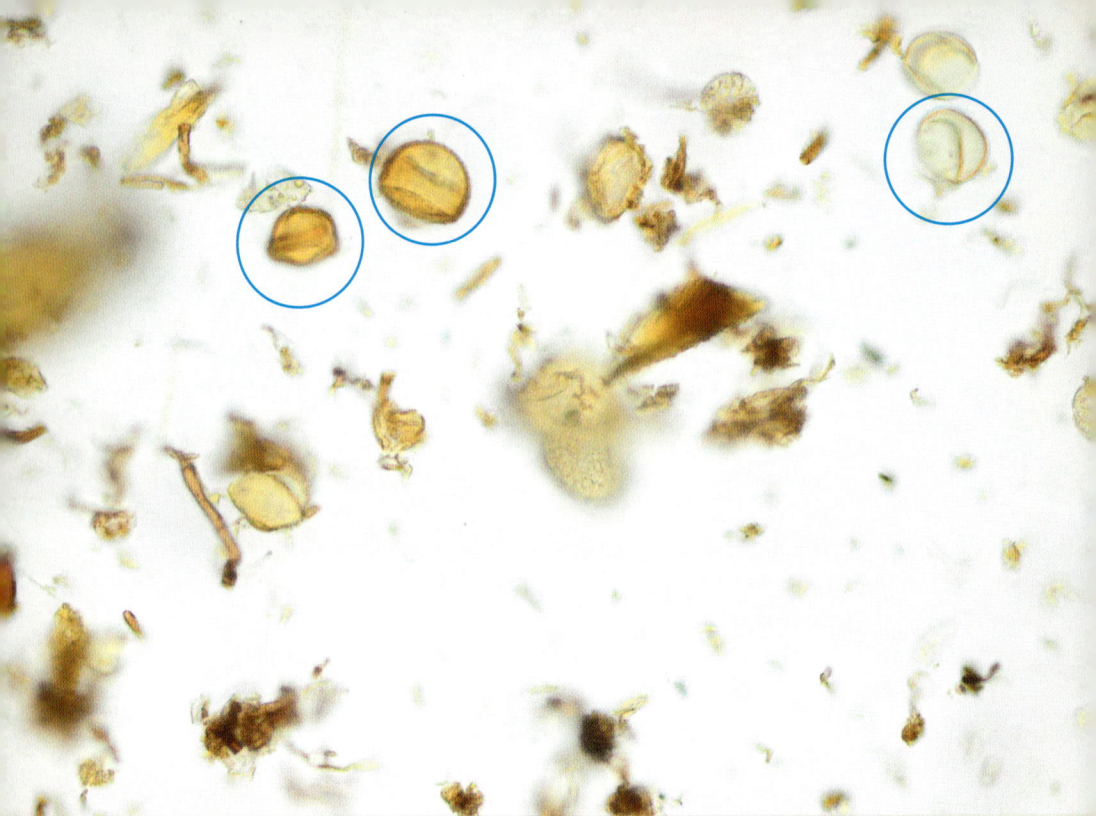

1:7 Peat deposits contain countless pollen grains (examples circled) that hold the record of Irish environmental change in the past

glacier. The river that flows today through the busy city once drained a great meadow of grasses and dwarf shrubs. During the short springs and summers of those times the little shrubby willow, with its scallop-edged leaves, grew in many parts of the country. Throughout the British Isles today dwarf willows live only in a few places where modern climatic conditions resemble those at the end of the last glaciation.

In some waterlogged conditions ancient seeds and leaf fragments may be so well preserved that they appear almost fresh. From ancient lake mud can be sieved the shiny, brown seeds of the bogbean (*Menyanthes trifoliata*) that once flourished on the edges of large lakes now long dried up. Plant parts particularly resistant to decay, such as hazelnut shells, form layers in peaty soils where hazel (*Corylus avellana*) bushes once grew. Occasionally, in bogs, long-dead pine seeds are found within shrivelled cones, dyed dark brown by the peat that has hidden them for thousands of years.

But of all the parts of plants that have been scrutinised for evidence of the past, by far the most prolific and most intensively investigated are the pollen grains and spores of the flowering plants, the conifers, ferns and mosses. Identifying the plants through their fossilised pollen grains or spores is reminiscent of studying the shells on a beach. The animals that once lived in the shells are long dead, the soft parts have rotted, but one can name the animal solely from the shell's characteristics. So it is with fossil pollen grains and spores that have lain long in peat and mud.

1:8 Alder catkins produce copious pollen that is distributed by the wind

The living material originally inside the pollen grain or spore died and decayed within a short time of the grain's release from its parent plant but the outer pollen grain or spore wall remained, being formed from a natural complex polymer, which is very resistant to bacterial and fungal decay.

The circumstances that bring pollen to ancient deposits are linked to plant sexual activity. Pollen grains are the male reproductive cells of flowering plants and conifers. They are these plants' equivalent of spermatozoa but, unlike sperm, pollen cannot move by itself; wind or insects may transport pollen grains. Plants that are wind-pollinated release huge amounts of pollen because their pollination mechanisms are hit-and-miss.

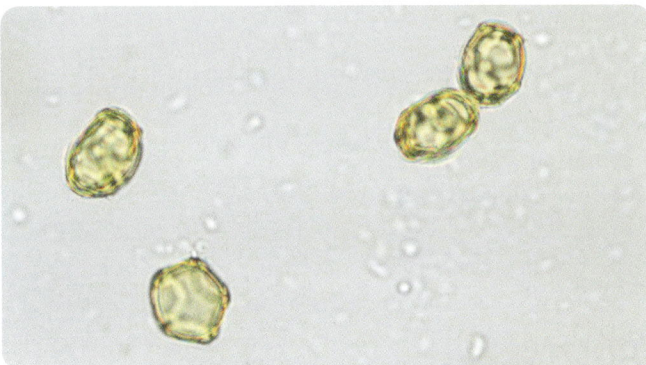

1:9 Light micrograph of alder pollen

1:10 The blossom of the blackthorn is pollinated by insects

Much less pollen is produced by plants dependent on insects for pollination as the insects that visit the flowers move the pollen from one flower to another. Although the pollen grain's purpose is to pollinate flowers of its own species, the majority never do so. Much of the dust in the summer air is composed of dead pollen grains and those that land on the surface of a bog or lake become incorporated into the deposit and fossilised.

Individual pollen grains are too small to be seen with the naked eye and are measured in thousands of a millimetre, a unit called a micron (μ). Pollen grains from Irish native plants usually fit the size range approximately 10–90μ. Buttercup plants are small and oak trees may be very large but their pollen grains are almost exactly the same size, approximately 25μ long. Unlike their sizes, the shape and surface features of pollen grains or spores vary greatly. The pollen grains from holly,

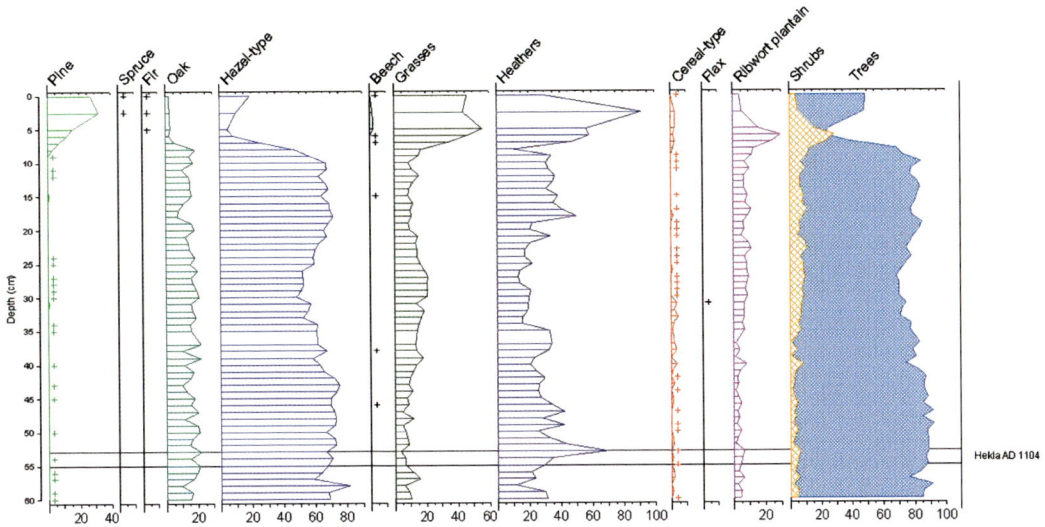

1:11 An example of a simplified pollen diagram representing changes in the behaviour of selected plant species through time. The pollen profile is from peat that accumulated over the last 1,000 years at Clonenagh Bog in County Laois Bog. A layer of tephra from the eruption of the Icelandic volcano Hekla in AD 1104 was detected in the peat profile and its depth in the profile is marked on the diagram.

for example, appear to be covered in little clubs, making them as distinctive as the tree's well-known dark green, spiny leaves. By matching the shapes and surface textures of fossil pollen grains with those from named living plants, the species that produced the fossil pollen grains can be identified. As pollen from plants that grew kilometres away may be blown onto the surface of the bog, there to be trapped, the suite of known sub-fossil pollen grains contributes to the detailed study of the vegetation that once surrounded the bog or lake.

The study of fossilised pollen and spores is known as pollen analysis or palynology and represents a powerful tool for the past environmental specialist. The pollen analytical method allows numerical studies of the pollen grains preserved in peat or lake mud to be handled statistically and shown graphically in 'pollen diagrams', whose interpretation allows past vegetation types to be reconstructed.

Pollen analysis was first developed in Sweden by a botanist, Professor Lagerheim, later refined by Lennard von Post, of the Geological Survey of Sweden and introduced to Ireland in the 1920s and 1930s the Danish scientists, especially Knud Jessen, working on tracing Irish woodland development since the end of the last Ice Age. The spread of vast birch and oak woodlands and later the subsequent spread of agriculture may be traced by pollen analytical investigations.

OTHER TYPES OF SMALL FOSSIL

There are other plant-like organisms, in lakes in particular, whose sub-fossils may be used to trace environmental change. These organisms are the algae and they are related to seaweeds. One type of microscopic alga has a single-celled body encased within two halves of a glass-like shell. These are called diatoms and the highly ornamented surfaces of their glass-like coats make them identifiable to species level. Certain species of diatoms live only in shallow, well-lit water and in such huge numbers that, over the years, their accumulated dead outer cases form a crumbly deposit called diatomite.

In layers some centimetres thick within the peat on the northern shores of Lough Neagh, whitish diatomite, locally known as 'Bann clay', may be found. These diatomite and peat 'sandwiches' are very rare deposits in Ireland and they have a strange story to tell. Peat forms on land but diatomite forms in water. The alternating bands of diatomite and peat indicate that in former times, perhaps about 4,000 years ago, the level of the lake varied greatly. Within the peat at Ballymacombs More, hundreds of metres west of the modern Lough Neagh shoreline, there are diatomite deposits formed when part or all of the lake was more extensive than it is today.

1:12 Diatoms thrive in fresh-water in such abundance that they form a crumbly white deposit

TRACING THE ANIMAL RECORD

The rather small body of evidence for Ireland's ancient fauna contrasts markedly with the wealth of evidence for Ireland's past vegetation. Interpreting the sparse faunal record needs a different approach from that used when dealing with plants. Unlike plants, animals may move over considerable distances, so fossil evidence from the place where an animal died provides only a limited amount of information about its life and habits. Many bones never survived long enough to get into the fossil record as dead bodies and marrow-rich bones were never safe from scavengers. Other factors also influence the fossil bone evidence. The wet, acidic conditions in bogs, so favourable to plant preservation, are less kind to bones.

Where soils are rich in limestone bones may remain in excellent condition and from these places comes the best evidence for a large, once-native mammal. The bones of the now extinct Great Irish deer (*Megaloceros giganteus*), lifted from the bottom of Ballybetagh Bog in County Wicklow (amongst other places), provide the richest evidence from any part of Europe for this impressive animal. The stags' skulls bear the largest antlers of any deer that has ever lived. The antlers and skeletons are all that remain of the big herds that lived in Ireland during a period of improved climate close to the time when the glaciation ended.

Other deposits contain rare reindeer (*Tarandus rangifer*) remains. Unlike the Great Irish deer, reindeer are not globally extinct although they are no longer Irish

1:13 The ribs of this human body, long buried in peat, have become as pliable as leather

natives. But reindeer lived in Ireland in the past, possibly in herds, grazing on short, scrubby vegetation similar to that in their modern native range above the Arctic Circle. Caves are good sources for the bones of Ireland's former native mammals, as cave deposits have contained bones of lemming (*Lemus lemus*), stoat (*Mustela erminea*), wolf (*Canis lupus*) and brown bear (*Ursus arctos*).

The fossil record is, however, almost silent on smaller creatures such as birds or lizards. It may be that, because their tiny bones are fragile, these animals' remains barely survived but further detailed examination of bone-bearing deposits may yield more minute bones to reinforce what is known of the smaller native vertebrates. Fish fare a little better than the birds or reptiles. The evidence points to a few freshwater fish species, including salmon (*Salmo salar*), trout (*Salmo trutta*) and Arctic charr (*Salvelinus alpinus*), which colonised lakes and rivers from the sea at the end of the last glaciation. The time of their first arrival in Irish waters remains unclear. The chaotic state of the Atlantic Ocean as the last Ice Age ended may have delayed their passage.

In contrast to the fish record, that for the molluscs is quite good. Shells of snails that live in freshwater, as well as marine molluscs such as oysters (*Ostrea edulis*) and mussels (*Mytilus edulis*), survive well. It is a lack of study rather than a lack of fossilised material that limits our knowledge of Ireland's native molluscs and the conditions under which they have lived throughout the last 13,000 years approximately.

Ancient scraps of beetles provide the most precise and detailed record of the swift, past climate changes experienced in Ireland and northwest Europe as the glaciation ended. Beetles breed quickly so they can respond to changes in temperature more rapidly than plants, with some liking warmth whilst others demand cool conditions. Beetles, like the other members of the huge insect group, have an outer skeleton made of a material called chitin. Chitin is almost as resistant to microbial decay as the polymer that makes the outer coat of pollen grains and, like pollen grains, beetles can be identified to species by comparing the characteristics of their fossilised bodies with those of living specimens.

THE CLOCK IN THE CARBON ATOMS

Within the bodies of all of the dead organisms that are of such value to environmental studies of the past there are further secrets locked up in the atoms that comprise the dead material. All living organisms contain carbon and it is small variations in this element's atomic structure that allows its use as a natural clock. Carbon atoms may take one of three physical forms known as isotopes, one that is radioactive and two that are not. The radioactive form of carbon is known as radiocarbon or carbon 14 and is expressed as ^{14}C. Radioactive carbon is extremely

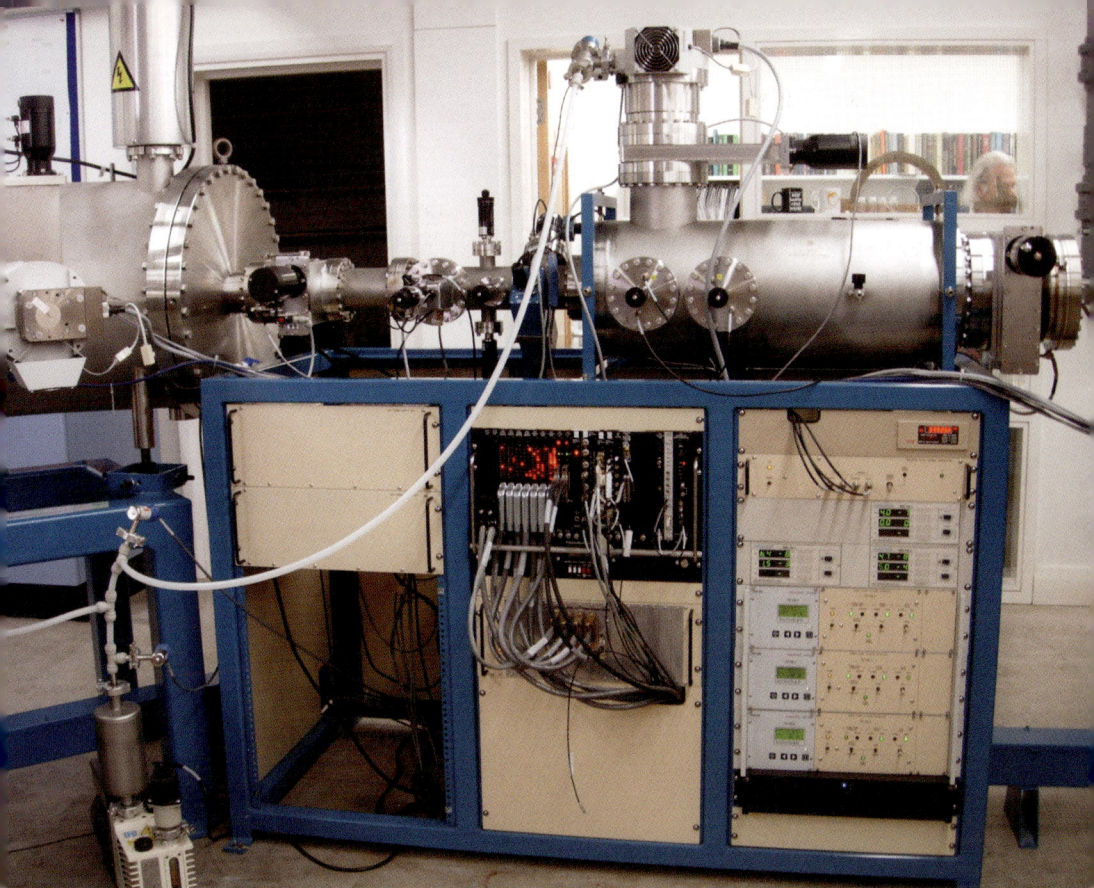

1:14 Part of the Accelerator Mass Spectrometer in the [14]CHRONO Centre, Queen's University Belfast, which is used to radiocarbon date a wide range of organic material

rare as only about one trillionth part of all modern carbon is [14]C. It is because radiocarbon decays over time, emitting radioactive particles as it does so, that it can be used as to date dead material, the principle being that the longer the material has been dead the less radiocarbon will remain.

Radiocarbon forms continually in the upper atmosphere by the interaction of cosmic rays with nitrogen atoms. The newly formed radiocarbon atoms join with oxygen atoms to form carbon dioxide gas (CO_2), which is chemically indistinguishable from carbon dioxide containing either of the other stable carbon isotopes. The minute trace of CO_2 containing the radiocarbon mixes through the atmosphere, eventually entering plants as they take up carbon dioxide during photosynthesis. Radiocarbon then enters the bodies of animals, first through the vegetable food eaten by the herbivores and then into the carnivores as they consume the flesh of the herbivores.

When a plant or animal dies, however, it no longer takes in radiocarbon and its corpse begins to lose the radiocarbon it had when alive so that, after 5,730 years, only half of the original level of the radioactive carbon remains. Therefore the 'half-life' of radioactive carbon is 5,730 years. If the level of radiocarbon remaining in a dead material such as seeds, bone, peat, wood, rope and leather is calculated then the age when the material was alive can be worked out.

How, therefore, can the amount of the remaining radiocarbon be expressed as years? If the rate of formation of radiocarbon had never varied, and knowing that it takes 5,730 years for half of the radiocarbon atoms to decay, then simple arithmetic could be used to work out when the seed or bone died. The situation is not that simple because the rate of formation of the radiocarbon has fluctuated through time and we cannot calculate the variation.

This apparent obstacle can be overcome using the oak tree-ring calendar, since ancient oak wood can be dated to the year and it too contains radiocarbon. To make the radiocarbon 'clock', blocks of oak wood of known age were taken, working back from the present time in twenty-year and, more recently, ten-year slices for almost 10,000 years, and the exact amount of radiocarbon in each wood slice was measured. To date dead material the amount of radiocarbon within it is compared with the values from the dated oak slices. The point at which the values from the dated wood match those from the dead material allows its age to be established. Radiocarbon dating is not as precise as tree-ring dating but it is a very valuable tool. The science of radiocarbon dating began in the 1950s and much of the research that utilised tree rings to turn radiocarbon estimations into real years was pioneered in Irish universities, especially Queen's University, Belfast.

There are a number of ways in which dates may be expressed. Dates of material older than two thousand years may be expressed as years BC, for example 2500 BC. Dates from the last two thousand years are expressed as AD, for example AD 1350. If a date has been derived through radiocarbon calibration this will be made clear. For example, the calibrated radiocarbon dates for seeds from a Bronze Age archaeological site may be expressed as 2000 cal. BC. An alternative to the BC/AD timescale is to express dates as years before present – years BP. There is a difference between 'present' and 'modern'. Modern moves forward in time so dates expressed as BP would be confounded because the modern dates of thirty years ago are not in alignment with those of today, whenever 'today' may be. Therefore, by international convention, 'present' is fixed at AD 1950. To convert approximately a BC to a BP date, add 1950, for example, 2500 BC becomes 4450 BP.

Throughout the text of this volume I have expressed dates using all of the above conventions and also used the term 'years ago' as this has a familiarity that is well suited to a book aimed primarily at the non-specialist reader.

THE VOLCANIC 'CLOCK'

There is a further natural calendar that is built into ancient deposits in Ireland. Tiny particles of volcanic ash from volcanoes that have erupted almost entirely in Iceland are carried to Ireland by weather systems that originate in the Icelandic region. There are only small traces of volcanic ash or tephra in Irish bogland and,

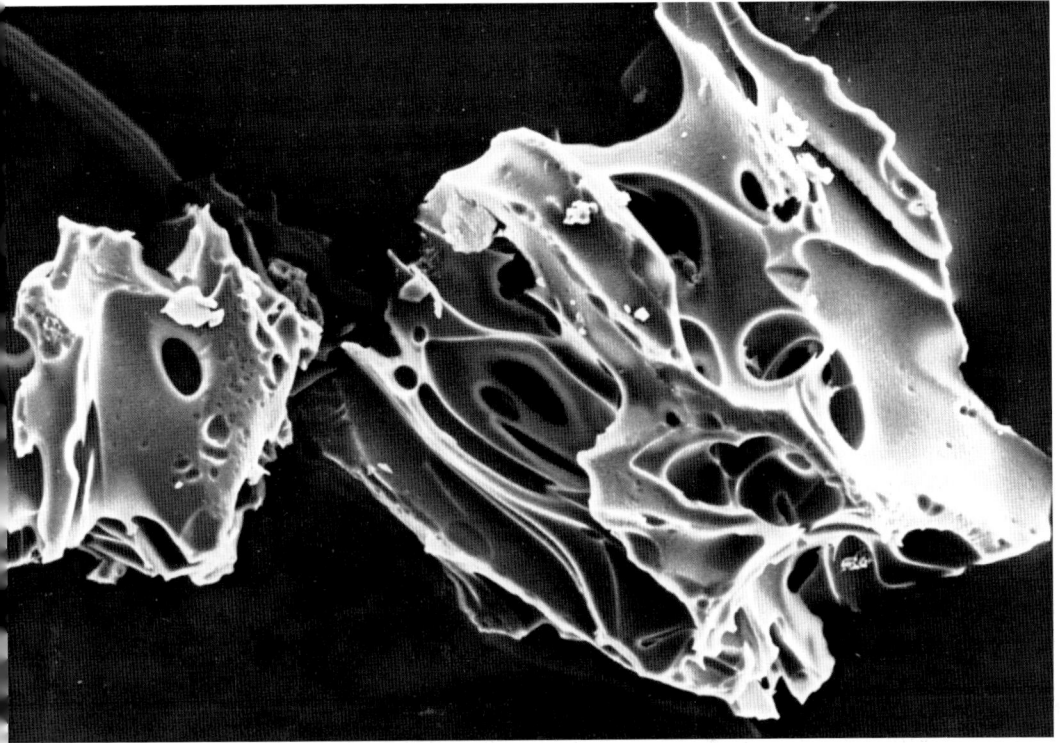

1:15 An electron micrograph of a tephra shard from a prehistoric eruption of an Icelandic volcano

because tephra shards are small, a microscope is needed to see these minute slivers of natural, bubbly glass. The shards of tephra from an individual eruption may be chemically unique and if the chemistry of a tephra in a volcanic ash layer is known and the layer dated, any sediment in which the geochemically unique tephra is discovered thus dates the material in which it is embedded.

Tephra layers from eruptions in Iceland that happened over the last 1,000 years were recorded in Iceland. People have lived there since the island was colonised in the ninth century AD. The numerous eruptions of the volcano Hekla in southern Iceland are amongst the largest of the last 1,000 years with the eruption in AD 1104 being noteworthy. Similarly the eruption of Öræfajökull in AD 1362, with other less violent eruptions recorded for dates throughout the last millennium. By matching the chemistry of the tephra layers found in some Irish bogs with the chemistry of the ash from the dated eruptions, the age of recent peat can be established precisely. Radiocarbon dating of an organic matrix is used to provide the age of tephra layers from eruptions that happened before people settled Iceland. Tephra studies have made a major contribution to understanding landscape change at the end of the last glaciation as deposits laid down then may be low in organic materials, making radiocarbon dating problematic.

PICTURES FROM THE PAST

At times, during the latter part of the last glaciation, southern parts of France and Spain enjoyed a mild climate whilst much of northern Europe, including the northern part of Ireland, was under thick ice. The cold north was uninhabitable but in southern France and Spain people hunted animals and gathered plant food from the wild. Amongst the people of southern Europe were skilled artists who used various pigments to paint on the walls of caves. Their thrillingly beautiful paintings, made between 35,000 and 10,000 years ago, are amongst the earliest works of art. They drew thousands of images of the animals that shared their world and some paintings show animals now extinct. These images are priceless records of the appearance of animals that have otherwise left no trace but their fossilised bones. Without these drawings made by people thousands of years before the last glaciation ended, we would have no knowledge of the colour or texture of the coat of the Great Irish deer. The legacy of work left by these superlative artists enriches the culture of ancient Europe.

Information is thus gained through study of maps, writings and the first drawings left by our ancestors, as well as tiny fossilised pollen grains, old bog wood, dead mosses and beetle cases. In combination with other studies of fossil remains and with the calendar for change made by natural 'timepieces', whose size ranges from large tree trunks to the smallest entity capable of independent existence, the atom, this data is the foundation for the story that will unfold on the pages that follow.

CHAPTER 2

THE ICE AGE ENDS

13,800–11,500 years ago

KEY ISSUES:

climate changes as the Ice Age ends;

colonisation by plants and animals;

colonisation theories;

vegetation development;

the arrival of large and small mammals;

climate change and disruption to the North Atlantic conveyor.

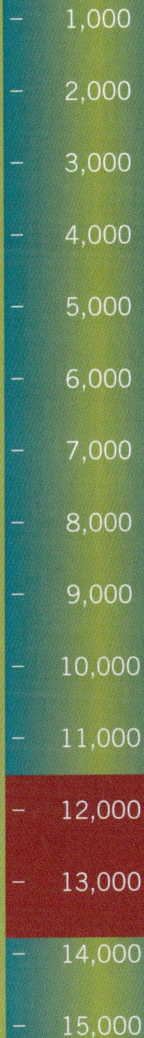

TIME PERIOD FOR THIS CHAPTER

YEARS AGO

- 1,000
- 2,000
- 3,000
- 4,000
- 5,000
- 6,000
- 7,000
- 8,000
- 9,000
- 10,000
- 11,000
- 12,000
- 13,000
- 14,000
- 15,000

THE END OF THE LAST ICE AGE

Twenty-one thousand years ago most of Ireland was covered by ice. Not a thin layer, like the one on the car bonnet on a winter's morning, but a heavy blanket more than 1 km thick in places. This massive ice sheet covered the north and most of the midlands. Only the land south of a line between Counties Limerick and Wicklow was probably bare of the ice covering and had possibly been so throughout the Ice Age that had lasted 120,000 years. It was during the period, around 21,000 years ago, that the Ice Age weather was at it worst. Even where the land was ice-free, little life could cope with conditions harsher than those at the Poles today. In those times the number of plant and animal species that lived in Ireland could have been counted on the fingers of a few hands. Occasional patches of red, green, brown or purple microscopic algae growing on the surface of snow may have been all that represented living organisms.

Conditions were about to change, however, for the Ice Age was coming to an end. The power of melting and moulding ice had a profound effect on the Irish landscape. Places that had been filled with ice were freed of its weight, with

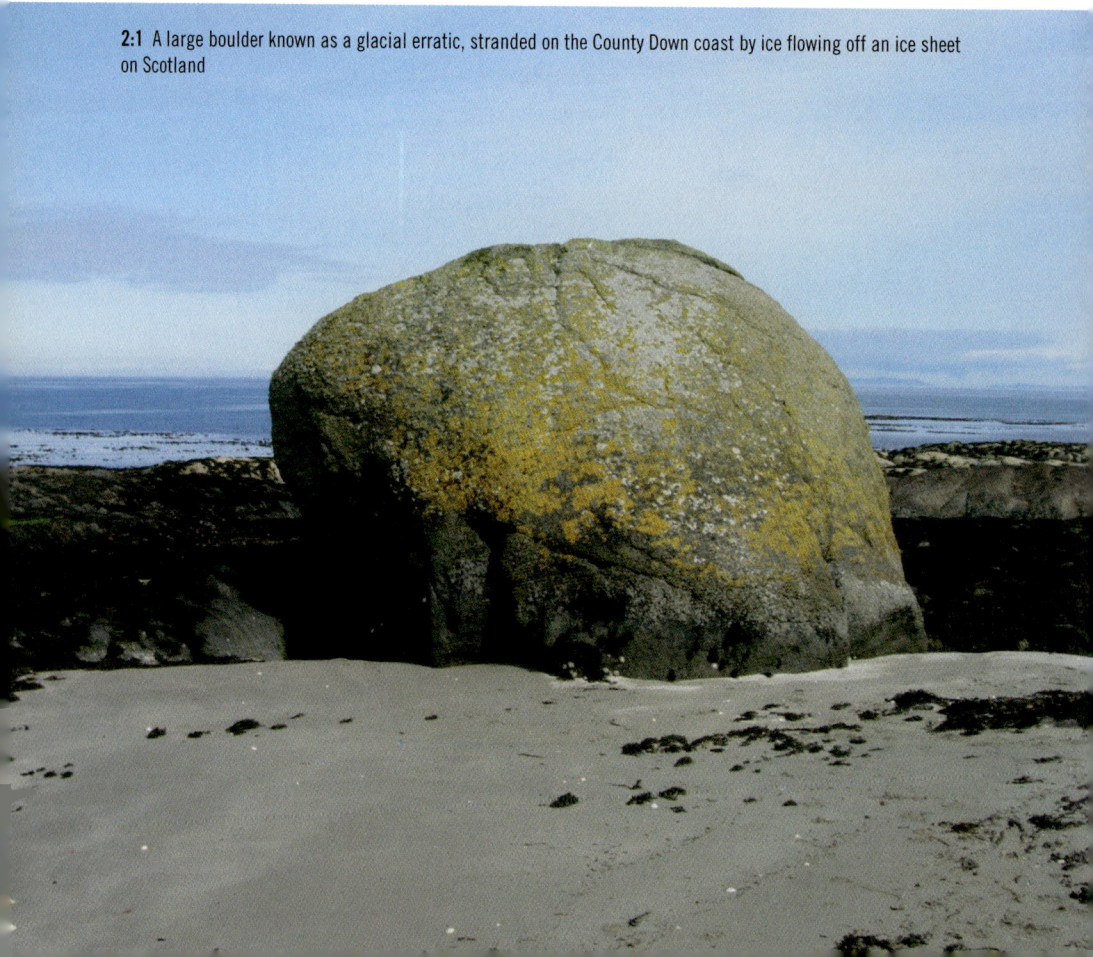

2:1 A large boulder known as a glacial erratic, stranded on the County Down coast by ice flowing off an ice sheet on Scotland

U-shaped valleys marking where there had once been glaciers. The River Shannon became established and the drumlins that dominate the northern landscape were formed as moving ice acted like a plough on the landscape. Large boulders known as glacial erratics marked places where powerful glaciers had dumped their loads of trapped rock.

The influences that can turn the Earth's climate from domination by cold weather to conditions like those of today are driven primarily by forces acting outside the planet. Earth's climate is controlled by the Sun so the position of the Earth in relation to the Sun controls the amount of radiation reaching the planet's surface. The orbital distance from the Earth to the Sun varies over a period of 108,000 years, as does the Earth's tilt to the Sun. The starting and stopping of Ice Ages are not only under the control of the relative positions of the Earth and Sun but these have a profound influence on the lengthy Ice Ages or glacials and the shorter warm periods, or interglacials, that separate them. We live in a time between two Ice Ages and have done so for over 11,000 years. If the timing of similar earlier periods remains a reliable guide, we have a few thousand years of pleasant weather remaining before the onset of the next Ice Age.

Twenty-one thousand years ago, when the climate was more severe than at any time during the last 125,000 years, so much of the Earth's water was locked up in the glaciers that covered vast areas of both hemispheres that global sea levels were about 120 m lower than they are at present. As the Ice Age ended, the huge glaciers melted and poured enormous volumes of water into the sea, causing its height to rise by about 8 mm each year. Sea levels rose steadily by about 55 m between 21,000 and 14,000 years ago, when the rapid melting of some of the global ice sheets released millions of litres of water into the sea every second for hundreds of years. The massive flow of water out of the southern hemisphere ice sheets forced water warmed at the Equator into the North Atlantic and helped to establish the Gulf Stream that has controlled Ireland's climate ever since.

As sea level rose and the ice diminished, the pressure on the land released and it began to rise. The coastline of the British Isles changed its shape and an archipelego of up to 6,000 islands formed, of which Ireland is the second largest. Ireland may have become an island sometime between 18,000 and 14,000 years ago when it is thought that the Irish Sea filled with water, but it took thousands more years before sea level rise separated many of the other British islands from the European land mass. Great Britain was not separated finally from Europe until just over 8,000 years ago (6100 cal. BC), probably by a huge tsunami that began off the coast of Norway. Proof that in late-glacial times a grassy plain had been where the North Sea is today comes from the fossilised mammoth bones and teeth, today trawled regularly from the North Sea's bed.

What record does the Irish landscape preserve of the changes that the land underwent as the last Ice Age ended? Amongst the mountains there are little lakes, once occupied by ice, and scratch marks on rocks scraped by stones trapped in the bottoms of glaciers. Occasionally on hillsides and beaches are large boulders, dumped as melting glacier retreated. Only in some of the highest places like Ben Bulben in County Sligo (530 m) and Slieve League in County Donegal (600 m) may the mountain tops have protruded above the ice sheet, forming isolated 'nunataks'.

THE RETURN OF THE PLANTS AND ANIMALS

As the weather warmed, plants and animals began to return and the evidence for their arrival lies beneath bogs and in lakes. At the bottom of the oldest bogs are often three layers of sediment. The lowest layer, which is over 14,000 years old, is mostly made of fine clay many metres thick, deposited by sediment-laden melt water from the ice. There are few fossils in this bottom layer, indicating that the Irish landscape was almost lifeless at that time.

Above this almost sterile layer is another that is often brown and muddy. It started to form about 13,800 years ago. At Sluggan Bog in County Antrim, the brown mud is made up of semi-fossilised microscopic water plants, bits of fibrous roots, fragments of *Sphagnum* moss leaves and parts of the plants growing nearby. At other sites the mud may also hold snail shells, bones, beetle wing cases and tiny pollen grains of plants, moss spores and freshwater algae. All of this partially fossilised plant and animal material shows that a wide variety of living organisms were doing well. This evidence for a richness of life contrasts with the layer above,

2:2 The stratigraphic sequence shown represents the basal, organic-poor clay, above which is a dark organic-rich layer, topped by a further layer of organic poor-material. The site is on the County Down coast at Roddansport.

which began forming about 12,700 years ago. In that layer there are fewer fossils, showing that living things were not as prolific as they had been previously. The layer is composed mostly of clay, silt and stones, along with a few thin layers of microscopic shards of Icelandic volcanic glass.

The colour and texture of the layers indicates that each formed under different conditions. About 13,800 years ago there began a short period of warming climate known as an interstadial. The first evidence for this warming in Ireland was detected at Woodgrange in County Down, hence it is known in Ireland as the Woodrange Interstadial. Improving conditions let plants and animals flourish for over 1,000 years, their dead remains producing the rotted material that made the muddy layer. Then, about 12,700 years ago, came a swift return to poor conditions known in Ireland as the Nahanagan Stadial (a stadial being a short period of deteriorating climatic conditions), once again making life difficult for many plants and animals for a further 800 years approximately.

HOW DID THE PLANTS AND ANIMALS GET TO IRELAND?

It is thought that few if any plants and animals may have survived in Ireland during the Ice Age and that the improving climate in the early years of the Woodgrange Interstadial led to changes that favoured the return of plants and animals. How did plants and animals get here, once conditions improved enough to support life? At the time when the Ice Age ended, sea levels were up to 150 m lower than today, doing away with the shallow seas that separate the British Isles from Europe in our own time. Greatly lowered sea levels left parts of the sea bed exposed and 'land-bridges' or 'walkways' formed for some plants and animals to cross from the continent. In the face of contradictory statements about the depth of the Irish Sea, it is asserted that ridges of glacial sand and gravel on the sea bed comprised the platforms and deep ridges that made land-bridges but computer models of glacial action do not confirm the land-bridge theory and there may have been an unbroken stretch of water in the Irish Sea as early as 16,000 years ago.

If colonisation was possible only by land, then the Irish Sea formed an insurmountable barrier but, cold and turbulent though it was, the sea could have formed a highway. Many animals can swim and parts of plants can float, with most unharmed by a brief time in salt water. Seeds and fruits washed down rivers and streams on the Great British side may have been swept across the narrow Irish Sea to be cast on the bare soils of the new Irish coasts. Floating plant parts may have carried small creatures, their eggs or larvae, and migrating birds could have carried seeds in their guts. The arguments for and against the processes that brought the pioneer plants and animals to Ireland are sure to continue as there is no single satisfactory explanation. Presumably the new arrivals followed a variety of routes, with rafting, floating, flying, swimming and possibly walking playing a part in bringing new life to Ireland.

There are about fifteen flowering plant species and some dozen invertebrate animals that are found in Ireland but not in Great Britain. The group is known as the Lusitanian or Hiberno-Cantabrian flora and fauna and is found in western Ireland as well as in parts of Spain and France. Some scientists assert that there were favoured places along the western coast of Ireland where plants species survived the worst climate of the last glaciation 21,000 years ago. Possibly the Hiberno-Cantabrian plants did survive under the better climate of the western coast, made less harsh by the warmth of the Gulf Stream, although at that time this ocean current was south of its modern range.

Pollen studies of occasional scraps of peat from earlier Irish interstadials consistently reveal a few species of water plant and three members of the heather family, St. Dabeoc's heath (*Daboecia cantabrica*), Mackay's heath (*Erica mackiana*) and Dorset heath (*E. ciliaris*). That these plant species survived in the earlier interstadials points to the possibility that they did so also when most of the country was under ice. These heather species have strong claim to being ancient members of the native Irish flora.

One or two of the modern Hiberno-Cantabrians may be deceivers. Studies of bog deposits close to the places where Hiberno-Cantabrians live today cast doubt on the native status of Irish heather (*Erica erigena*). The plant may have been introduced only about six hundred years ago as its pollen has been identified in peat dated to about AD 1430 or AD 1480 from Claggan Mountain and adjacent sites in County Mayo. There is a hint that it was brought by pilgrims. The plant grows near Bilboa in northern Spain, close to Santiago de Compostela, a major pilgrimage site. An alternative explanation may be that the plant was shipped in as animal bedding or packing material. In contrast, the strawberry tree (*Arbutus unedo*) or 'wolloghan tree', as it was once known, grows in the Kerry oak woods; although it has been said that it too was introduced during the medieval period, the evidence for its native status is strong, since examples of its wood, dated to about 4,000 years ago, have been found. The implication is that the plant is an Irish native species.

In spite of contention about the ways in which plants and animals arrived in Ireland, they came in increasing numbers as the climate of the interstadial became established. The interstadial lasted a little over a thousand years, during which the vegetation of some parts of the lowlands changed up to three times, partly in response to climate. Changing soil conditions, rainfall patterns, extended periods of cover by snow, frost action, wind strength, altitude, light levels and the effect of animals on plant populations also affected the establishing vegetation cover.

2:3 *Arbutus* is a rare native Irish shrub

PLANTS THAT COPED WITH CHALLENGING CONDITIONS

The soils of Ireland at the start of the interstadial were difficult for plant growth. Bare mineral soils become waterlogged in winter and bake dry as pottery under the summer sun but there are some plants that thrive in soils hardly worthy of the name. It was grasses (*Gramineae* species), sorrel dock (*Rumex* species) and members of the daisy family (Asteroidaceae) that brought the first leaves and flowers to much

2:4 Plants of modern wasteland are characteristic of vegetation at the end of the last Ice Age

of the country. Huge swathes of the landscape must have been smothered in the plants that farmers many millennia hence would call weeds.

Agricultural weeds have qualities that make them almost immortal and it was those same abilities that let these plants thrive in late-glacial soils. The landscape changed with the seasons for the first time in thousands of years as the docks, grasses and occasional ragwort (*Senecio jacobaea*) plants gained ground. Wearing a new covering of grasses, docks and ragwort, much of Ireland must have had the appearance of modern wasteland.

For the first few hundred springs the ground was heaped with the bleached stems and leaves of the previous year's grass, amongst which were the frilly green leaves of over-wintering ragwort rosettes. As the spring sun warmed the soil, reddish spears of dock stems grew quickly, so that by summer the spring landscape had been refreshed with new colours. Yellow ragwort flowers blossomed alongside the lime-green flowers of the docks. As the year progressed and the daylight hours shortened, the dock flowers formed upright heads of rusty-brown seeds. By the year's end, only the dead grasses, the dock seed-heads and the blackened, withered ragwort stems covered the landscape in advance of the winter snow.

The tiny dwarf willow, no higher than the adult human ankle, was once common throughout the lowlands but is today restricted to the tops of a few high mountains where conditions resemble those of late-glacial times. The dwarf birch lived in Ireland at that time and when both minute shrubs were doing well, each year as winter turned to spring, they covered the land in yellow catkins. Lake vegetation also did well and many rivers and pools had their surfaces striped by long, floating leaves and stems of pondweed.

It is the relative lack of trees in the early interstadial that marks Ireland's vegetation as different from that of Great Britain, where woodlands had begun to form. The reasons why woodland was so uncommon in Ireland are not clear. There were, however, a few places in Ireland where the vegetation was more like that of Great Britain: for example, in Killarney in County Kerry where birch (*Betula pubescens*) occasionally formed patches of woodland, as it did at some favoured sites just north of Lough Neagh and parts of Counties Donegal and Clare. At Woodgrange there is a tantalising glimpse of the start of woodland for here a few pollen grains of poplar (*Populus tremula*) were found in the lake mud and also the pollen of the sea buckthorn (*Hippophae rhamnoides*). Today orange-berried sea buckthorn is common on some Irish sand-dunes but it is no longer native. It was a rarity that died out at the end of the late-glacial, to be reintroduced to Ireland during the nineteenth century.

Seeds from deposit at Mapastown in County Louth show that here too the sea buckthorn grew on the dry soils and in the damp places there was scurvy grass (*Cochlearia* species) and sea pink (*Armeria maritima*). Today the sea pink is found growing along the coast and in the high hills, all that is left of this plant's earlier more widespread distribution. In County Donegal, around Lough Mullaghlahan and Altar Lough, there was patchy dock-strewn grassland. The wet west and north still had extensive areas of unstable soil, on which crowberry (*Empetrum nigrum*) heath did particularly well, since the plant can manage on soils that have lost nutrients to high rainfall.

THE FIRST ANIMALS

The diversity of the early late-glacial vegetation must have guaranteed an equally rich insect fauna. The insect fossil record is far from complete, though nothing like as silent as for the spiders. Insect and spider bodies are fragile and only a few insect types are found as fossils. Late-glacial lake deposits contain wing-cases of beetles (Coleoptera) and the head capsules of non-biting midges (Chironomids), which may be identifiable to species. In addition to revealing the diversity of native beetles, their response to climate change provides new insights into local conditions.

The record for the first mammals is reasonably good, even though fossilising conditions may not be ideal for preserving animal remains. In many places the chemistry of the acidic soils damages fossils that contain calcium. The early

2:5 A modern reindeer antler collected from the Arctic island of Svalbard. It was an antler like this that formed the fossil found in ancient lake mud on the coast of County Down.

grasslands were nutritious enough to feed reindeer. The fossil evidence for reindeer is very sparse, with only about twenty-six sites throughout Ireland having yielded reindeer remains. An intertidal late glacial deposit at Roddans on the County Down coast yielded fragments of reindeer antler, proving that these animals once lived where the sea is now.

Reindeer are well adapted for life in the cold, with their heavy coats of insulating under-fur and cloven hooves well suited to walking on soft ground and snow. Reindeer will eat whatever the changing seasons bring, browsing on buds and newly unfurled leaves in spring, then relying on lichens, twigs and mushrooms during the colder months. The lack of reindeer fossils in the north and west of the country suggests that crowberry was not their preferred food, though they could tolerate it.

It is unclear if the vegetation changes recorded at various places all happened at the same time or varied by decades. The climate had continued to improve, with conditions good enough to let birch (*Betula pubescens*), eared-willow (*Salix aurita*) and juniper (*Juniperus communis*) take hold. In a few places in the east and

2:6 Scarlet pimpernel can grow in poor, unstable soil

southwest, shrubs and trees formed thin woodland. In the northeast juniper was widespread, possibly due to its resistance to cold. Its needle-like leaves contain oils and resins that act as an antifreeze.

A brief worsening of the climate did not suit trees and shrubs, which began to disappear from the landscape and the wet weather opened soils to erosion. The presence of plants like mugworts (*Artemisia* species), chickweed (*Stellaria media*), scarlet pimpernel (*Anagallis arvensis*) and members of the pink (Caryophyllaceae) family that thrive in shifting soil shows in the pollen record.

The climate soon rallied and, once again, grasslands dominated the rich soils. Amongst the grasses were many spring- and summer-flowering plants, including yellow-flowered rock roses (*Helianthemum* species), silverweed (*Potentilla anserina*), with its buttercup-like flower and silver-backed leaves, and meadow rues (*Thalictrum* species), with others that brought vibrant colour to great expanses of waving grass stems.

THE LANDSCAPE OF THE GREAT IRISH DEER

Trees were no longer the only large living organisms on the Irish landscape. There were increasing numbers of the Great Irish deer (*Megaloceros giganteus*), feeding on luxuriant vegetation. The remains of this big, native mammal can be seen in museums throughout Ireland, as people have been finding giant deer bones for hundreds of years. For example, there is a sketch of an antlered skull unearthed in County Meath in 1588 that was subsequently sent to Hatfield House in Hertfordshire and mentioned in State Papers of Elizabeth I. Additionally, there is a documentary record made by Dr Thomas Molyneaux in 1697.

Blackened, fossilised skeletons and antlered skulls of the majestic Great Irish deer have been found in ancient lake mud throughout the lowlands. Twenty-eight skulls and three complete skeletons have been recovered from the Ballybetagh Bogs in County Dublin, the first of these found in 1847. The three small bogs at Ballybetagh hold the record for the maximum number of deer skeletons found in Ireland. Remains have also come from Downpatrick in County Down, Lough Gur in County Limerick, Whitechurch in County Waterford and Ballyoran Bog in County Cork but not the southwest. The Great Irish deer is a true deer, not an elk, although the animal is often known as such.

The Great Irish deer is not unique to Ireland, as its bones have been found in other parts of British Isles as well as Europe and Asia. The animal is called 'Irish' because its remains are found more commonly here than in any other part of its range. The skeletons of these huge deer are large and, when muscle, fat, skin and hair are taken into account, a fully grown male could have been almost 2 m high at the shoulder, 3 m in length and could have weighed 700 kg, with females a little

2:7 The head of a male Great Irish deer showing the huge antlers

smaller and without antlers. The Great Irish deer was long-legged and quite finely built, despite its huge size. It had comparatively long ears and a tail that flipped and curled.

The males' distinctive heads were ornamented with impressive antlers and it is these that confer on the Great Irish deer the distinction of having the largest antlers of any deer that has ever lived. The males' skulls had strengthened bones along the central line to stop the weight of each antler from splitting its head. The neck vertebrae have long, vertical spines and these, with the surrounding muscles and tendons of the neck and shoulders, formed the pronounced hump. No skins remain preserved to show how the animal looked when it was alive but there are Old Stone Age cave paintings in France that show the Great Irish deer with a dark shoulder hump, contrasting with the pale brown coat. The paintings also show the deer with large eyes, emphasised by a dark stripe and a broad, blunt muzzle.

The grasslands of Ireland at the end of the last Ice Age were the perfect environment for the herds of Great Irish deer that may have lived mostly in the lowlands, for fossils are not found in the uplands. The blunt muzzle gives a clue to the food eaten by the deer as its shape is ideal for closely cropping grass. Analysis of the microscopic scratches on Great Irish deer tooth enamel confirms the dietary clue from the muzzle shape, since the marks on the teeth are typical for an animal that fed primarily on grass. The Great Irish deer shed its antlers every year and, as each stag got older and the antlers got bigger, the animal needed more calcium and phosphorus in its diet. Antler growth rate was rapid throughout the 150 days of the spring and summer growing period.

Based on what is known about herd behaviour in modern deer, it may be that for most of the year male and female Great Irish deer lived in separate herds, the does occupying the grasslands on higher ground and the stags grazing on the damp pastures beside lakes. In late summer, as the mating season started, the sexes combined. Each stag occupied a territory. Stags roared and nodded their heads so that their huge, light-coloured antlers flashed in the autumn sunlight. While some stags were displaying their masculine charms for all the females to see, others were fighting by locking antlers, pushing and wrestling until one of the combatants submitted. At the time of the rut parts of the Irish lowlands must have been as noisy and active as the Serengeti Plain in Africa are today. When mating was complete, the males and females separated again. Older experienced females lead groups of about a dozen others back to higher ground, leaving the small herds of stags to descend to the valleys where they recuperated from the exhaustion brought about by fighting and energetic coupling.

At all the Irish sites where Great Irish deer skeletons occur, those of males greatly outnumber those of females. A number of theories have been put forward to explain why female skeletons are so rare, although a headless skeleton would be difficult to sex.

2:8 The female Great Irish deer has no antlers

It has been suggested that skulls of females have not been recognised, possibly being confused with those of horses. Alternatively, the female skulls may have been cast aside because they bore no antlers and would not fetch much money. It is probable that female skulls are genuinely very rare. The National Museum of Ireland has the largest collection of Great Irish deer bones and amongst the almost 200 skull and antler specimens only six represent females. In the meadows where the females spent most of their lives, there were few places where bones could fossilise. Perhaps the females' bodies left little trace because the bones were destroyed by scavengers.

The pleasant weather and lush vegetation of the mid late-glacial were about to change dramatically, as events on the other side of the Atlantic Ocean were poised to have a massive, catastrophic effect on the world's climate. Just before the catastrophe occurred the climate throughout the country had improved yet further, as it did throughout much of the North Atlantic area, even as far north as Greenland.

Then, as now, any alteration in the flow of the water of the North Atlantic Ocean affects the world's climate and the magnitude of the change that was about to happen can hardly be overestimated. Movement of the North Atlantic begins at the Equator, where the sun warms the surface water, which is then driven northward by the wind. The northward flow continues until the warmed water meets the cold

Arctic sea near the coast of Iceland. The Arctic waters cool the warm, dense, salty sea surface water, which then sinks down through the ocean until it reaches the bottom, all the while drawing more warm water from behind. When the cooled water reaches the ocean's depths, the direction of its flow reverses, therefore, as the surface waters are moving north, the deep waters travel south. This turning over of surface and deep water creates a conveyor belt, whose flow influences the movement of the water of all the oceans. Anything that disrupts the climate powerhouse that begins in the North Atlantic will eventually affect the climate of the Earth.

A RETURN OF COLD WEATHER AND ITS CONSEQUENCES

Events on the eastern North American continent 12,700 years ago were about to nearly destroy the North Atlantic conveyer. Much of the North American continent had been covered by two vast ice sheets. The central area of the vast expanse of eastern ice had been melting, producing a lake of melt water far larger than the modern Great Lakes. The enormous ancient lake was held back from the North Atlantic Ocean at the coast. When a breach occurred at about the place where the modern St Lawrence River flows today, vast amounts of freshwater and icebergs flooded into the North Atlantic. The gargantuan discharge of water and ice, greater than the flow of all of the modern rivers of the world combined, prevented the warm surface water from reaching its former northern limits so it was no longer able to meet the cold Arctic Sea. The conveyor became sluggish as the power it needed to function was greatly weakened and, in less than fifty years, the Earth's climate system crashed back to conditions resembling those of the preceding Ice Age, producing in Ireland the conditions of the Nahanagan Stadial. The winters worsened and the growing season shortened from 150 to 120 days, with the loss of growing days most crucially in early spring. At times summer temperatures may have been, on average, as much as 12°C lower than in the preceding interstadial.

Breaks in the vegetation appeared when rain, snow, frost and wind caused soil disturbance and erosion. Where soils were disturbed, as at Tory Hill in County Limerick, mugwort (*Artemisia* species), meadow rues (*Thalictrum* species) and knotgrass (*Polygonum* species) invaded the sedgy grassland to form an Irish version of tundra. The dwarf willow must have coped better than many shrubs because its fossilised leaves turn up at numerous sites. Along the western coast, from the Dingle peninsula in County Kerry in the south to Donegal in the north, crowberry heath managed to hang on, being tolerant of the wet and the cold.

As vegetation deteriorated, the animals began to suffer and life became difficult for the herds of Great Irish deer. The soils lost nutrients to rain, melting snow and surface run-off. Soil nutrients were already depleted because overgrazing of the

2:9 The pollan is a native Irish fish

grasslands by herds of reindeer and giant deer during the interstadial had robbed them of calcium and phosphorus. And, as the deer were heavy animals, their weight on vegetation that was already fragmenting helped to destroy further the essential grazing lands.

The shortened growing season and the dirth of good quality food took its toll on the herds. Animals could not lay down adequate fat reserves for winter so there were deaths during the cold months. Time passed and the herds dwindled, as more deer died than were born. In a few hundred years the Great Irish deer was extinct, having lived in Ireland for less than two thousand years. The herds of reindeer had also died out, leaving the landscape to the tiny, sharp-toothed Arctic lemmings (*Dicrostonyx torquatus*) that could eat the stems of sedges and grasses and the lean-bodied stoats (*Mustela erminea*) that hunted them. The rivers and lakes were still inhabited by a few species of native fish, notably the pollan (*Corgonus autumnalis*), possibly the first fish to colonise Irish lake waters, as well as Arctic charr, salmon and trout.

Sea buckthorn and dwarf birch became extinct and none of the deer species survived to become members of the community of mammals that made it through to the improved conditions of the interglacial to come. The plant and animal species that survived the stadial were soon to swell greatly in number as new plant and animal species arrived at the onset of the interglacial.

CHAPTER 3

NEW PLANTS AND ANIMALS

11,500–9,500 years ago

KEY ISSUES:

the meaning of native status;

plants on the coasts;

the arrival of the trees;

early mammals.

TIME PERIOD FOR THIS CHAPTER

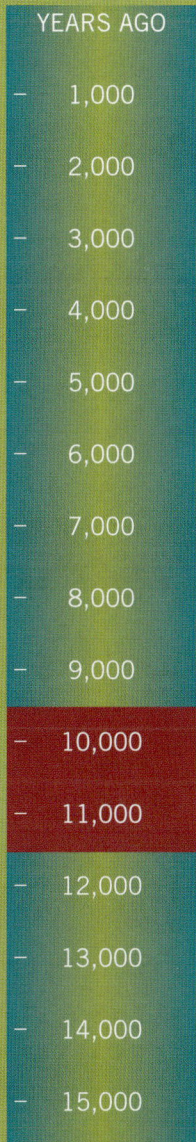

YEARS AGO
1,000
2,000
3,000
4,000
5,000
6,000
7,000
8,000
9,000
10,000
11,000
12,000
13,000
14,000
15,000

EVIDENCE FOR NEW CONDITIONS

Above the layers holding the evidence for change as the Ice Age ended are many metres of peat that have been accumulating for approximately 11,500 years. These peats are the deposits that formed during the post-glacial period. A sustained period of improved climate between Ice Ages lasting thousands of years is known as an interglacial, and we live in the current one, known as the Holocene Interglacial and also referred to as the post-glacial period. The term Holocene is derived from Greek, meaning 'entirely recent'. The term was first coined in the mid-nineteenth century and it was then adopted for deposits that are wholly recent in their fossil content, for example, Irish peat and lake mud.

The sharp transition as the clay gave way to Holocene peat makes it plain that, in Ireland, the interglacial began swiftly – there is no fossil evidence for a gentle, prolonged period of improving weather.

3:1 Two examples of the sharp transition between the deposits of the end of the last Ice Age and those of the present interglacial

Everything points to the climatic improvement happening over decades rather than centuries and resulting in conditions similar to those of today. It is not known precisely when the Holocene's first influences were experienced in Ireland but in Greenland the climatic change is marked in the ice layers that accumulated 11,700 years ago.

Within the swamp peat that grew at the edge of lakes as the Ice Age ended lies the evidence for profound landscape and climate change. Boglands are an unrivalled source of information about Ireland's past landscape, with further information coming from additional sources such as ancient shorelines, where fossilised remains of drowned woodlands along the Irish coast show that sea levels have varied through time. Ancient animal bones from places like Castlepook Cave in County Cork provide evidence about some of the first animals to return to Ireland as the post-glacial period began.

In Ireland the first years of the Holocene saw improving conditions and the fossil record shows that new species of plant and animal arrived and colonised the warming land and sea. The fossil record for the plants and animals is not perfectly synchronised, making it difficult to disentangle the story of which came first and radiocarbon dating is not sufficiently precise to separate the short, swift phases of the earliest colonisation. For the purposes of this account, a general assumption will be made that the country was colonised in phases and that, as new plant species arrived, this was mirrored by an increase in the number of new animal species. In response to the newly warmed conditions, throughout the country much of the new vegetation began to establish in a few decades to a few hundred years.

There was no lack of variety in Irish plant and animal life during the preceding stadial but only a limited number of species survived its harsh conditions to reappear during the early years of the Holocene. Juniper may have been the only major woody species left from the previous climatic regime, continuing to grow as the soils began to warm. Organisms that survived the difficult conditions of the stadial represent the small founder population that was much enlarged by the arrival of many new species. The evidence is strongly supportive of the colonists coming from places beyond Ireland. These colonisers are now deemed to be true natives of Ireland, leading to the question 'what are the criteria that a plant or animal must satisfy to be worthy of Irish native status'? It is notable that only a few plant species may be unique to Ireland and even these are considered dubious by some botanists.

WHAT IS A NATIVE IRISH PLANT OR ANIMAL?

To be considered an Irish native, a species must have arrived in Ireland before there were people. The interglacial started sometime after 11,700 years ago and the first people began to arrive as early as 8,000 cal. BC, so in the intervening period, and following the strict criteria for native status, all such species, from the unnamed micro-organisms living in the soil and water to the largest and best known trees, had to arrive and establish.

Native status may depend on early radiocarbon dates for bones, insect bodies, seeds and wood; fossil evidence that proves that the animal or plant lived and died in the country before the arrival of people and, recently, DNA studies have contributed to the native status debate. Native status cannot be attributed to a plant if only its pollen grains or spores are found because these may have travelled considerable distances on the wind. For example, the wood of the beech tree (*Fagus sylvatica*) has never been found in Irish peat or lake mud that formed before the arrival of people, even though its pollen grains are occasionally found in these deposits. These pollen grains probably came from beech trees growing in the southern English woodlands.

If one adheres strictly to the criteria, then the question of which plant and animal species are true Irish natives is impossible to answer conclusively in every instance. There has been a tendency to assume that if a species known to be native to Great Britain is also found in Ireland, then it must also be native to Ireland. It is not known how many of Ireland's supposed natives have become 'native by association' with those on the neighbouring island. At present there are about 850 seed-producing plants thought to be native to Ireland and some 150 species of ferns, about 67 per cent of which are also native to Great Britain. The larger, neighbouring island boasts about 1,000 seed-producing species and is home to about 4,000 species of beetles, whereas Ireland's native beetle species may number about 1,000. Ireland has about 400 native species of spider, whereas approximately 650 are native to Great Britain. Of the amphibians and mammals, in Ireland there are no common bank voles (*Clethrionomys glareolus*), moles (*Talpa europaea*), common toads (*Bufo bufo*), crested newts (*Triturus cristatus*), palmate newts (*Triturus helveticus*) or snakes such as the grass snake (*Natrix natrix*).

3:2 The grass snake is not a native Irish species

3:3 Salt marshes such as this began to develop during the early Holocene

Where did the new natives come from? Great Britain and Europe are the most likely places from which plants and animals expanded their territories, partly in response to better weather during the growing and mating seasons. A European source should not be discounted on the basis of greater distances. Studies of the DNA from Irish pine marten, pygmy shrew, stoat and hare show these species have a DNA signature more like those of their continental relatives than those in nearby Great Britain. As the Holocene began, the same mechanisms that controlled colonisation at the start of the interstadial appear to have been active again, this time involving even larger numbers of new plant and animal species. The contentions that surround Holocene colonisation are much the same as those during the Woodgrange Interstadial that have been considered in the previous chapter.

The early years of the Holocene saw a new type of vegetation taking hold, firstly in the lowlands, for the broadleaved woodlands were on the way. Imagining the appearance of the early post glacial landscape of Ireland faces the reader with an intellectual challenge. Modern Ireland contains little that resembles the very early landscape and it is not easy for the mind to grasp the many swift changes through which the land was about to pass.

3:4 Members of the goosefoot family grow where small pockets of water-holding material have accumulated

CHANGES IN THE LANDSCAPE

To begin with, the Ireland of 11,000 years ago was not the same shape as modern Ireland. Sea levels were still rising, resulting in changes to the coastal outline. The zone between high and low tides was colonised by seaweeds, marine worms and molluscs. Above high tide, where soils were sandy, dunes began to build, whilst in the estuaries, the first salt marshes began to form.

All of these changes have had to be surmised from studies of early coastlines, as the plants and animals that may have lived in these new habitats have left virtually no fossil record. Along many modern Irish beaches there are clues to the ways in which the first sandy and stony areas became colonised by plants. At the top of the beach the tide piles heaps of seaweeds wrenched from rocks below the low-tide mark. The seaweeds dry and their fragments lodge between the stones and gravel,

3:5 Wild iris grew luxuriantly on the edges of lakes and ponds

helping scarlet pimpernel, orache (*Atriplex* sp.), sow-thistle (*Sonchus* sp.) and cleavers (*Galium aparine*) to germinate and root.

These plants do well in stony places that have some source of freshwater and nutrients. In sandy spots, grasses and members of the Umbellifer family of plants, of which wild carrot (*Daucus carrota*) is typical, take advantage of open soil. All of these plants can thrive because they are prolific producers of seed, even where animals use them as sources of food.

Inland, the lakes and rivers saw the arrival of the native fish, especially pollan, Arctic charr, salmon and brown trout. If only more was known about the early migratory history of the eel (*Anguilla anguilla*) in Ireland. White water lilies (*Nymphaea alba*) grew in the still shallows of lakes and wild iris (*Iris pseudacorus*) did well on the marshy edges.

At the bottom of many bogs there are fragments of reed stems that, although they are thousands of years old, look as fresh and pliant as reeds only recently dead. In lake mud the outer skeletons of midge larvae show that lake bottoms provided good food for insects. We may assume that in those early times also, the adult midges swarmed during spring as they do at Lough Derg today.

The luxuriant plant growth at the edges of shallow water produced masses of leafy material. Throughout each autumn and winter the dead leaves and stems of water lilies, false bulrushes (*Typha* sp.) and water mint (*Mentha aquatica*) collapsed into the water, to rot and lay down a layer of fibrous mud, ultimately filling the damp hollows. Willows thrive in damp ground and were next to colonise the wet edges, spreading so that the former lake grew into a wood. In these places mosses did well, with *Sphagnum* growing on the moist floor of the new woodland. The length of time taken for a shallow lake to turn into a wood varied greatly but between tens and hundreds of years could elapse before open water was turned to woodland.

Along some shorelines, to the north and east of the country in particular, the rising level of the sea prevented freshwater from draining away from the depressions in the sand-dunes. At the back of the dunes shallow pools of freshwater formed and in these, as in the inland pools and lakes, plants lived and died and their remains formed the first coastal peat. The crowberry heaths of the west dwindled as trees began to take hold of the land.

THE ARRIVAL OF THE TREES

Almost everything known about the arrival of the trees to Ireland comes from pollen analytical studies. These can give a general indication of the relative proportions of the various types of tree in the landscape. All the tree species that came to the British Isles appear to have done so from Europe. The trees that founded the new Irish forests could have come from many places on the continent, where they had survived in favoured refugia throughout the Ice Age. Trees like oak and ash survived in places as far apart as Spain and the Balkans and, to the north, lived Scots pine as well various species of birch and willow. It is entirely possible that each new species invading Ireland may have done so from more than one refugium. Recent studies on oak tree genetics show that the modern Kerry and Fermanagh oaks originate from different places.

Left to colonise only by growing into new territory, the trees would have taken many thousands of years to do so, progressing at less than half a kilometre a year. The extremely swift spread of the new forest trees must have been helped through dispersal of their seeds by birds, including those that were migratory. The spread of oak was probably helped by jays (*Garrulus glandarius*), rooks (*Corvus frugilegus*)

3:6 A copse of birch in winter. Such copses were common throughout the lowlands as the tree spread throughout the country.

and wood pigeons (*Columbaria palumbus*). Alder's woody fruits could have been spread by water or bird species. The insects that depend on the trees for a living may have crossed the sea on the driftwood coming out of forests bordering rivers in Great Britain and Europe. The partially rotten wood was ideal material on which eggs and larvae may have 'hitched a ride'.

It is not warmth alone, however, that allows trees to spread into new localities. Conditions for good growth also depend greatly on soils, as these have a profound influence on plant colonisation. Where soils were under-developed, tree spread must have been slowed. Trees already into the Irish midlands were still missing from the north, as the soils there were not mature enough to sustain the new arrivals. Birch, for example, is susceptible to drought and needs soils that are sufficiently mature to hold water, otherwise the tree's seeds do not germinate or the seedlings do not establish. In other areas, sandy or gravelly soils or extreme altitude may also have hampered birch's expansion.

The deciduous birch soon expanded its range to cover much of the Irish lowlands. Its swift arrival and rapid spread, almost as soon as the Holocene started, hints that the population of birch whose seeds arrived in Ireland may not have been very far away. It has been suggested that birch trees survived the previous cold period in sheltered places on the Dingle Peninsula in County Kerry and on the land beyond. As a pioneer birch is ideal because it set seeds at an early age and does so reliably year by year. Birch seeds are light and readily spread by the wind and the tree grows well on a range of soils. Soon after its arrival the dark hues of the evergreen juniper gave way to the new birch-dominated landscape, interspersed with willows and poplars. Massive tracts of the lowlands and extensive areas of the lower parts of the mountains were covered by birch wood.

The seasons were marked across the country by almost synchronous changes of leaf colour. In the late winter, the thin branches of the birches were hung briefly with yellow catkins, giving way to a short period in mid-spring, when all the young leaves unfurled, greening the landscape within days. During the summer period the cool, stable air and protection from sunlight within the woodland suited insects and other invertebrates. As the year progressed into the autumn, a further simultaneous change in leaf colour turned the woodland canopy to yellow and, when the growing season ended, the falling leaves returned the trees' appearance to silvery bark.

Each year the birch woodlands shed huge amounts of leaves. Of all the native trees, birch has associated with it the largest range of fungal species, some of which helped to rot the leaves whilst others rotted the fallen wood. Much of the tonnage of nutrient-rich, fallen birch leaves decayed and contributed to the humus accumulating in the soil. As the organic content of the soil increased, earthworms, insects and microscopic organisms multiplied, further preparing the soil for the next wave of newcomers. Next to arrive was the Scots pine, firstly along the southern coasts, accompanied in some places by yew. Juniper, Scots pine and yew are Ireland's only native conifers. Scots pine is tolerant of a wide range of soils and manages to thrive under difficult weather conditions.

Once other trees with thick, leafy canopies established, the birch was doomed as it cannot withstand shade. From the pollen records it would appear that about five hundred years after establishment, the birch woods were greatly depleted because of the invasion of shade-producing hazel.

The question of how hazel could have intruded into forests of birch must be considered. The watery conditions in the lowlands may provide the answer. The central lowlands were a world of lakes, pools, streams and shallow plains that flooded every winter and hazel did well in the soils at the edges of the lakes. Anything that landed on the flood-plain soils could have been distributed many kilometres as the waters rose in the autumn and dropped again each spring.

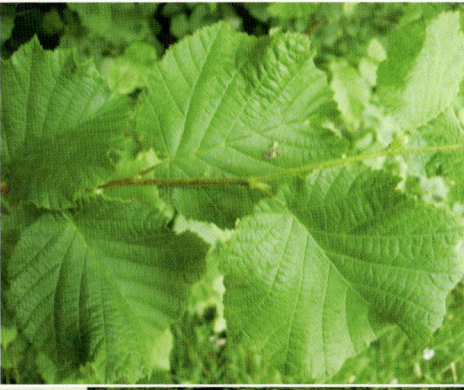

3:7 The large leaves of hazel shade any plant beneath them

3:8 Modern hazel takes the form of a shrub but it can also assume the proportions of a tree

Under good light conditions hazel produces large quantities of nuts with the ability to float, and so these could have washed into the lowland birch woods by winter flooding, to germinate in the rich, silty soils left after the water fell. Even if only a portion of the nuts germinated, the hazel would have grown up through the birch, assuming the dimensions of a tree and casting shade, with each succeeding generation of hazel relentlessly crippling the birch. The young hazels would have been successful because birch seedlings could not grow under the shade of hazel, whereas hazel seedlings could thrive under birch. The large quantities of fossilised hazel pollen found in peat and lake mud that formed at this time may hint at sunnier skies under which most hazelnuts could grow to maturity. In Ireland today cloudy skies dominate and it is only the well-lit bushes that can produce quite modest crops of nuts. Through the proliferation of hazel we must consider the possibility that the skies of the past may have had less cloud cover.

The hazel woods looked completely different from those of the birch. Birch trees, with their fine branches and small leaves, contrasted in shape with the solid, multi-stemmed hazel shrubs covered in large, floppy leaves. Today, hazel usually forms a shrub, although it can grow into a tree.

As each winter turned to spring, the bare hazel branches sprouted female flowers, looking like tiny tufts of carmine thread. The male catkins released great clouds of pollen so that the sunlit air was thick with yellow motes. Pollen fertilised the female flowers, thus starting the processes that would lead to pale green nuts forming on the outer twigs, often in pairs or trios, each surrounded by a ruff of

45

leaf-like bracts. The nuts, which are 2–3 cm long, may have grown singly or in bunches of up to seven nuts, darkening as they ripened, turning glossy and brown by late autumn and then falling to cover the ground, along with the leaves. The leaves begin to turn to yellow in late August and September but remain on the shrub longer than those of most other deciduous native trees.

Hundreds of thousands of tons of protein-rich nuts, up to 0.5 tonnes per hectare, were produced each year, with much of this bounty feeding animals in variety. The damage caused to the birch woods by the hazel would soon be reinforced by shade from elm and oak. Through its shade alone oak can eradicate birch from woodland in about 150 years. Oak and elm take time to establish but once they do so there is no getting rid of them. They cast shade because the arrangement of their leaves and branches lets each tree make good use of light.

The regional diversity of the new woodlands was quite marked and there is no single account that applies equally to the whole country. The first place where hazel prevailed was in the Dingle Peninsula, in County Kerry, where it may have grown unchecked by competition from other trees, in contrast to the Burren in County Clare, where pine and hazel arrived almost at the same time. From County Donegal in the northwest to eastern County Antrim the landscape may have changed in as little as one hundred years, from vast acreages of birch woodland to a sea of hazel. In the northeast especially, almost pure hazel woods grew for the next six hundred years. Hazel did well in the midlands also, especially on the warm, nutritious soils of the eskers, the big ridges of gravelly soils left by the retreated glaciers. The record for hazel in the uplands is sparse, since the early soils that may have preserved its remains eroded millennia ago. Today hazel is found above 600 m in the north of Scotland and one may deduce that in Ireland's past it may have formed open woodland in places now devoid of trees.

In the far west the landscape was not as hazel-smothered as it had been previously by birch. Oak, pine, elm, and yew were present as rarities and, where conditions allowed, birch, poplar and willow held out. Where there were no trees in the wet exposed places, fragments of crowberry heath lingered. To the far southwest the woodlands of the Dingle Peninsula in County Kerry were different from those in the north, west and midlands as birch and hazel occurred together and, in the scrubby places, docks grew. It must be assumed that birch and hazel grew side by side rather than in a mixture that included occasional oaks, pines and elms. Along the banks of the lakes in County Cork willows grew on the moist soils. Less is known about the southeast but on the sandy soils of dry southeast Wexford, pine was scarce but oak was more plentiful.

Other trees such as alder and ash were present and would, over time, dominate parts of the lowlands, but not in the early days. In the first woodlands that sprang up and spread, there were further trees and shrubs that have always played lesser roles. Hawthorn (*Crataegus monogyna*), wild cherry (*Prunus avium*), wild apple

3:9 Hawthorn, blossom (above) and berries (below), grew along woodland edges

(*Malus sylvestris*), and wild roses (*Rosa* sp.) for example, all of which have white or pink-tinged blossom and red fruits, lived on the woodland edges that were also home to many birds.

There are over three hundred native Irish birds and it is likely that the blackbird (*Turdus merula*), the song thrush (*Turdus philomelos*), with its brown-speckled, cream breast, and the red-breasted robin (*Erithacus rubecula*) sang then as they do in woodland today. The swallows (*Hirundo rustica*) that had returned as migrants from the south as the climate warmed, swooped on abundant insects. At one time the quiet of the Irish woodlands was disturbed by the hammering of great spotted woodpeckers (*Dendrocopus major*) and there were capercaillies (*Tetrao urogallus*), a large member of the grouse family that lived on the floor of pine woods. The modern absence of both of these birds is most likely due to woodland loss, particularly over the last couple of centuries. It is thought that the long-eared owl (*Asio utus*) has lived in Ireland for millennia but not the tawny owl (*Strix aluco*).

The beech (*Fagus sylvatica*), with its characteristic grey bark and glossy green leaves, is not a native of Ireland although, with the small-leaved lime (*Tilia cordata*) and the hornbeam (*Carpinus betulus*), it grew well in southern England, where it was a major native forest tree. Its absence from the list of native Irish trees, based on there being no finds of its wood, has been attributed to it not having crossed the southern land-bridge before this was overwhelmed by rising seas. If, however, colonisation did not depend on dry land, then a different reason for the absence of the three British natives from Ireland must be considered. Lime and hornbeam are grown in Ireland today but they do not set seed, so if they did arrive at the start of the Holocene they may not have been able to establish, even though temperatures are thought to have been higher at that time. Today beech is planted throughout the country and its seeds germinate freely in open, disturbed ground. It is this aspect of its life history that may help to explain why it never gained a 'root hold'. If the soils beneath the early woodlands were not suitable for germinating beech seeds, it would not have established well enough to have spread and successfully competed with the other tall trees.

EARLY MAMMALS

If the reason for the lack of beech in the early woodlands is perplexing, then the history of the native mammals is even more so. The large mammals, such as the Great Irish deer and the reindeer, disappeared from the fossil record before the Holocene began and, as yet, only a little is known about the fortunes of the smaller, furred animals like the wood mouse (*Apodemus sylvaticus*). Recent studies, in which radiocarbon dating has been put to good use, indicate that there is almost no fossil bone evidence for any mammals and little for freshwater fish or amphibians in Ireland before the arrival of the first people. On the other hand, hare, stoat, pine marten and pygmy shrew have DNA signatures more like those of their relatives in Europe than Great Britain, increasing the possibility that these mammals are truly native.

Although the current suite of evidence points to the early landscape having no mammals, the story for the amphibians is not as clear-cut. Whether or not the

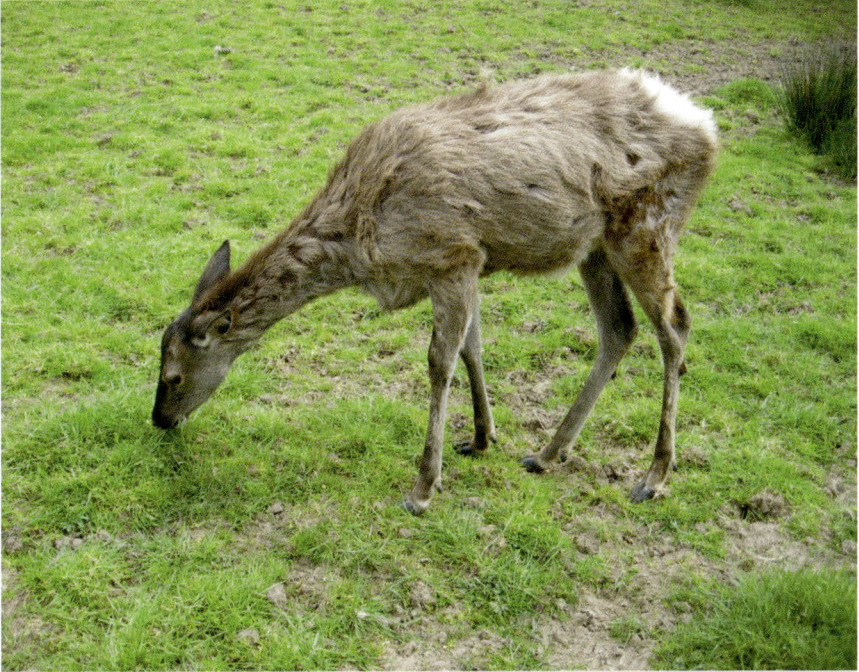

3:10 The native status of the red deer continues to be debated

natterjack toad is native has been a topic of argument by naturalists investigating Ireland's past. Studies of the DNA of the natterjack toads (*Bufo calamita*) that live in County Kerry show it to be different from those in neighbouring Europe. This difference could have happened only if the Irish toad population had been isolated for many thousands of years. These new findings give some support to the argument that the noisy, little amphibians have been living in Kerry for most of the Holocene.

We are, therefore, forced to consider that the wolf (*Canis lupus*), forest-dwelling wild pig (*Sus scrofa*) and, probably at a later period, red deer (*Cervus elaphus*) were brought into Ireland by people. The earliest known occurrence of red deer in Ireland is from sites in Belfast, at Ormeau and Sydenham, where bones dated to about 7,000 years ago have been found but here is much contention about this date.

Archaeological investigations have shown that some of these animals were brought to the Scottish islands in prehistory by people, presumably as very young animals, and this information indicates that they could have arrived in Ireland in the same manner. The apparent absence of the better known mammals from the early Irish record does, however, provoke thought. There is an almost complete dirth of sites with animal bones that have no hint of human influence but the absence of evidence is not absolute proof of the genuine absence of these animals. Over the last five and half millennia, changes in the landscape and the activities of

3:11 The badger is better known dead than alive in Ireland today

3:12 The skeleton of an aurochs. These large forest cattle were not native to Ireland.

people have wiped out the scraps of evidence that may have helped to clarify the status of many land animals. It is sad that today most of our familiarity with the country's wild mammals comes from the bodies of those killed by traffic.

Returning to the larger mammals. Red deer can live in and on the edges of woodland and their presence may influence woodland structure because they eat tree seedlings growing in well-lit clearings and bark from the trunks of trees. Wild pigs plough the soil with their powerful snouts, keeping it well churned. Trees stripped of their bark may die if the inner tissues are damaged and wounds on trunks or roots let in fungi and insects that can kill. But there is nothing to show that deer or wild pigs were present in the earliest Irish woodlands, so does this mean that these woodlands were more closed than those in Great Britain and nearby Europe?

In contrast to Ireland, in Europe and in Great Britain gaps in woodland were kept open by big grazing animals such as aurochsen (*Bos primigenius*), the original wild cattle of the woods, and by red deer.

In Europe there were herds of deer, wild cattle and European bison (*Bos bonasus*), whose influence on woodland structure is thought to have been significant. By eating seedlings and saplings, these big mammals kept the clearings open. In Ireland, however, there are no fossil bones of aurochsen and, as we have discovered, the record for red deer is slender and contentious. Surprisingly, therefore, comparisons of pollen records from Irish, European and Great British early woodlands show that all had similar structures, whether there were big animals present or not.

In the densest woods old trees died and eventually fell, letting light reach the younger trees below. Lightning and the fires that followed killed healthy trees and strong wind toppled them. In sunny clearings the old and decaying trees provided kindling to feed fires as well as homes for insects that lived on rotting wood, for example the long-horn beetle (*Anaglypus mysticus*). In an old woodland there may be as many dead trees as those still living. Seasonal flooding of woods close to rivers enriched soils and maintained the open spaces needed by the light-loving, smaller plants. It is the fossil record from the little animals however, such as some of the land snails and insects with their requirements for light or dappled shade, that may be better than pollen studies at revealing the mosaic of habitats in the early Irish woodlands.

During approximately the first two thousand years of the post-glacial period, the whole country went though a series of changes that turned the open, treeless landscape into one that had largely become a forest; but the landscape was not a sea of trees from coast to coast. There is much to show that in the woodland clearings, on lake and coastal edges and in the high places there were open areas where light-loving animals and plants could find a home. The docks and light-loving grasses were now restricted to places where trees did not cast killing shade. The open limestone landscapes of the west and, in all probability, the slopes of the treeless uplands provided refuge for plants and animals that could cope with increasing warmth but not with lessening light. It is, however, the great woodlands that overwhelmed the lowlands and progressed into the higher places that characterise Ireland at that time. And it was to this wooded landscape that the first people came to Ireland.

CHAPTER 4

THE ENVIRONMENT OF THE FIRST IRISH PEOPLE

9,800–6,000 years ago

TIME PERIOD FOR THIS CHAPTER

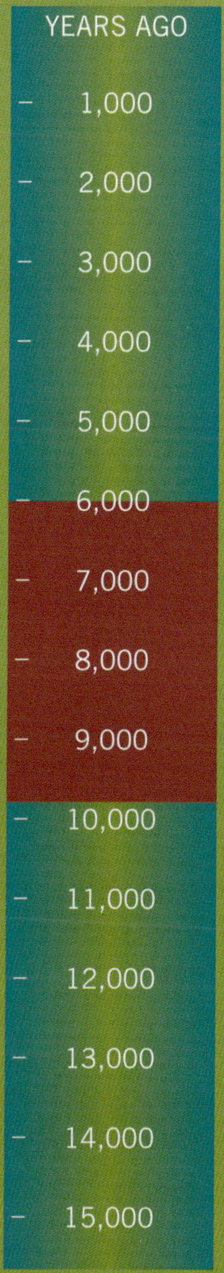
YEARS AGO
1,000
2,000
3,000
4,000
5,000
6,000
7,000
8,000
9,000
10,000
11,000
12,000
13,000
14,000
15,000

KEY ISSUES:

bog forests;

lives of the earliest settlers;

the rise in alder.

THE GREAT BOG WOODLANDS

When exploring the lowland raised bogs from which turf has been cut from vertical faces to provide fuel, we are faced with dilemmas. 'Turf' has been cut and dried for hundreds of years and, in the process, large areas of Irish bogland have been damaged beyond repair but the cuttings also provide insights into former landscapes. Embedded near the bottom of the deepest cuttings is wood, so soft that it disintegrates when taken from the oldest peat.

Above the fossil birch and willow, there may be a layer of roots and trunks of bog pine, which in contrast to the willow and birch wood is hard, for bog pine does not waterlog.

Occasionally, beneath a big pine trunk there are needles as well as cones that may be over 8,000 years old. The cones' seeds are long dead but their scales still open and close in response to the humidity of the air, just like their modern counterparts.

Right: **4:1** The light-coloured material is ancient wood, so soft that it was easily penetrated by the metal coring device

Below: **4:2** Bog pine timber protrudes from a lowland peat bog cutting

Many of the extensive lowland bogs began to develop 9,500 years ago, although some such as Sluggan Bog in County Antrim are older still. As well as holding bits of damp wood, the peat also envelops the pollen grains of the vegetation that grew on the drier high ground nearby. Wood from the dry land forests does not remain but the trees' pollen grains blew onto the young bog surface to be captured in the peat. Therefore, by joining the evidence from the wood in the peat and pollen from nearby woodlands, a picture of the forested landscape of over 9,000 years ago may emerge.

THE FIRST BOG WOODLANDS

The Irish climate entered a period of reduced rainfall about 8,300 years ago. Lessening rainfall may have coincided with changes in bog hydrology but the result was that the bogland water tables fell and bog surfaces dried enough to let pine seeds germinate. Before this time pine trees had been restricted to the dry land but the species is quick to colonise any place it finds favourable. Pine cannot grow with its roots permanently wet but the tree can thrive in conditions where nutrients are not plentiful.

In the Sluggan Bog peat a densely packed layer of fossilised timber gives the impression of a great pine wood. Tree-ring dating showed that the trees began to spread onto the bog surface in 6361 BC. When the bog timbers were dated, it emerged that during the 800 years that followed there were three periods when bog pine woodland expanded onto the bog surface. The first trees grew large and strong but eventually became the victims of their own success. The roots of the pines sucked water from the bog whilst the heavy trees compressed the peat, thereby altering the hydrology of the bog. Climatic records show that the climate then became wetter and a subsequent rise in the bog's water table made the peat so wet that the trees died. The bog underwent a further drying phase before the next generation of pine seeds germinated, demonstrating that the woodland's growth phases were linked to climatic factors as well as being forced by the hydrology of the bog.

It is interesting to note that where the trunks protrude from the cuttings, many lie pointing in the same direction. The roots of bog pines had fused together and if one tree fell, others were also loosened. Wind-throw after big storms could have toppled trees so that the trunks ended up lying in one direction on the bog surface, soon to be buried in the accumulating peat.

Soon after the first of the bog pine forests became established, the weather worsened. About 8,200 years ago most of the North Atlantic region experienced a period of cold and wet that lasted about forty to fifty years. This, the coldest phase of weather since the Ice Age ended, may have been the result of the final collapse

4:3 The roots of a modern Scots pine tree cross and fuse, like those of pine that once lived on ancient bog surfaces

4:4 Modern oak seedlings can thrive on peat for a few years but they will die eventually through being starved of nutrients

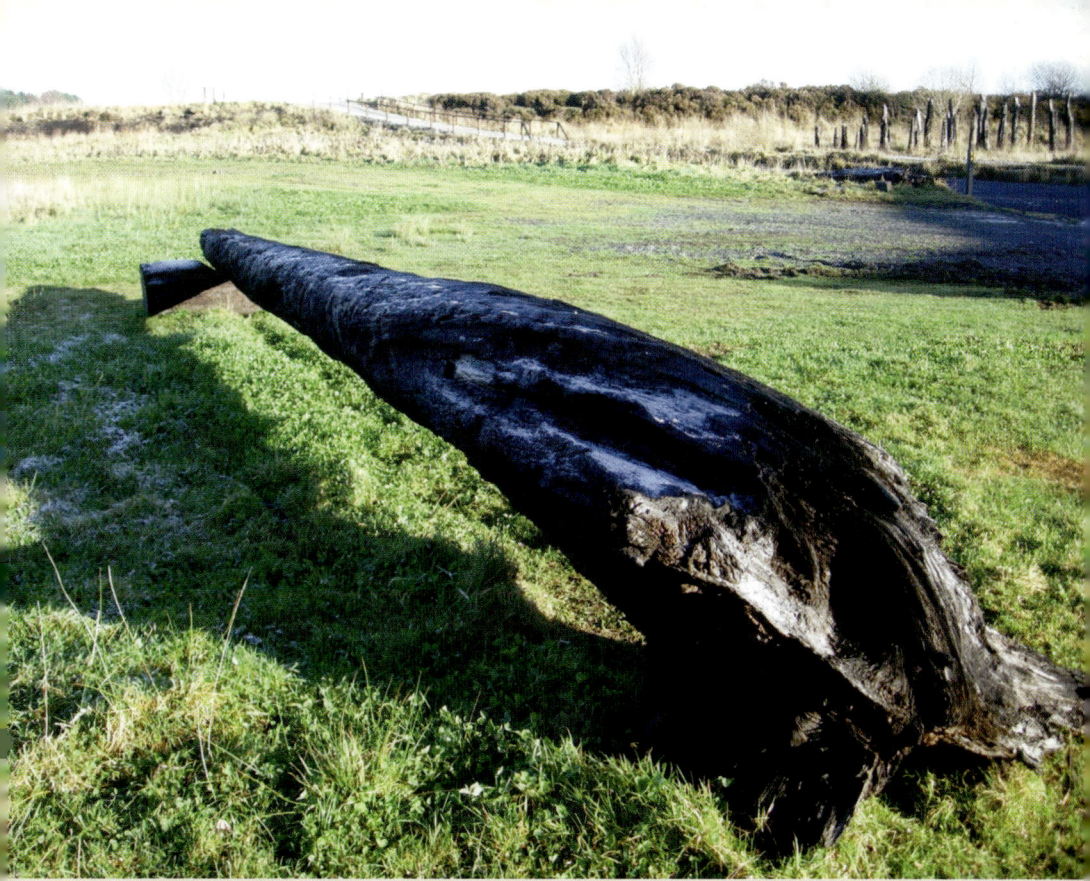
4:5 A large bog oak trunk, once embedded in a County Offaly peat bog

4:6 One of the oldest tree trunks in Ireland, the trunk exposed on the stony beach at the Roddans, County Down.

of the ice sheet that had covered much of eastern Canada, allowing a final flush of freshwater into the North Atlantic.

The next tree that found a new niche on bog surfaces was oak. Over 7,000 years ago, conditions on the edges of some fens became suitable for oaks. Bog oaks demanded more nutrients than pine and therefore the trees were restricted to the edges of mineral-rich fens and, in some places, the trees rooted into mineral soils or clay lenses (or layers) underneath the peat, thereby benefiting from slightly better feeding. The majority of the bog oaks, of which hundreds of thousands have been dragged out of peat to litter fields and drains throughout the country, date from the period 7,200 to 2,200 years ago (5200–200 BC), although these ancient trees are not the oldest oaks found in Ireland. A few trees that have been radiocarbon dated are almost one thousand years older. One tree trunk found on the stony beach at the Roddans in County Down dates to 6572–6432 cal. BC and from County Waterford came a tree dated to 6471–6356 BC.

The main body of dendrochronologically–dated bog oak samples are from the area north of Lough Neagh. These, with others from Ballymacombs More in County Londonderry and Garry Bog in north County Antrim, all of similar age, indicate that the bog oak woodlands started in 5200 BC and then increased after 5100 BC.

Even when conditions for the bog oaks were at their kindest, the trees were growing under duress. At times when worsening climate made the peat colder or wetter, the oaks endured but put on little wood. A slice though a bog oak trunk will show the rings that grew reliably, year by year. From time to time very poor growing conditions are registered by bands of extremely narrow rings, some marking major climatic events. The more exotic of these include the influences of past huge volcanic eruptions on Irish climate. Even though many of the bog oaks grew under demanding conditions, they had an average lifespan of 220 years. A tree excavated from a peat bog in Drumalure Beg in County Cavan holds the longevity record of 428 years, having lived from 4249–3822 BC.

Bog forests did not grow on all bogs. The upland peat on the borders of Counties Leitrim, Cavan and Fermanagh are almost free of bog pine and bog oak. In contrast, Ballymacombs More in County Londonderry contains thousands of bog pines as well as oaks and there are places in County Kerry where bog oaks are quite numerous. Research indicates that the distribution of the bog oak and pine woods were primarily under the control of conditions on the bogland.

DRY LAND WOODLANDS

Large and varied forests grew almost everywhere, stretching from the dense woodlands of the lowlands to lighter cover on the mountain slopes of MacGillycuddy's Reeks, the Nephin Begs, the Silvermine Mountains and other high places. Most of the tree species that grew in the lowlands also grew at greater altitude, with the mountain ash (*Sorbus aucuparia*) growing where it was too cold for most other trees. Sadly, little is known in detail about the primeval mountain woods or the altitudes that they reached.

Soil types could have a profound influence on the tree species that made up the woodlands. The limey soils of the midlands were suited to elm as this tree demands good feeding, whereas oak or hazel grew without much regard to soil type. The density of the woods also varied across the country. In County Westmeath there were woods dominated by elm, mixed with pure stands of hazel trees and with oak a rarity, in contrast to the lighter mixed woodlands of Donegal. Further west long-established pine and birch continued to dominate.

Plants that grow underneath the trees have their growth and flowering controlled by light penetrating from above. During summer the leafy canopies cast shade on the woodland floor, so ferns, mosses and liverworts, plants that could cope with low light levels, formed a green carpet between forest giants. During the winter and spring, however, when all the trees' leaves had been shed, light reached the floor of the woodland, favouring plants that needed light to bring them into flower. These plants had to complete their flowering in the brief period between the time when the sun's warmth took the winter cold out of the soil and the point at which shade intensified a few weeks later. In some woodlands there may have been thousands of yellow primroses (*Primula vulgaris*) and in others bluebells (*Hyacinthoides non-scripta*) or white-blossomed wood anemones (*Anemone nemorosa*), grasses and sedges, all flowering profusely.

The edges of the woodland were also well lit and there the twining stems of honeysuckle (*Lonicera periclymenum*) and ivy (*Hedera helix*) scrambled up the tree trunks to take advantage of light. The honeysuckle's golden flowers bloomed high in the branches in summer whereas the greenish flowers of the ivy opened during the winter, providing food for insects. Beneath the trees the ground may have been over-run with thorny blackberry (*Rubus fruticosus* agg.) stems. Fruits of wild roses, cherries and elder (*Sambucus niger*) fed animals and eventually the people who would soon begin to colonise the country.

LAKES AND COASTS

In the lowlands the striking features were the huge lakes and mighty rivers. The River Shannon ran a 200-mile course from its complex source that includes Cuilcagh Mountain in County Fermanagh and the Shannon Pot in County Cavan, through the midlands and on to the sea. In the early Holocene its waters were already carrying the load of sediment to the south that settled and eventually formed the soils of the wet riverside meadows or 'callows', as they are known in Ireland. Lough Neagh, Lough Derg and the lakes of the midlands were much enlarged in those ancient times, especially during the winter when their flood plains were filled to capacity, providing food for thousands of over-wintering water fowl.

In counties Offaly, Tipperary, Galway, Clare, Roscommon, Westmeath and Cork there is an indication that the lakes were also larger than they are today. The evidence comes from curiously shaped stones found beyond the edges of the modern lakes. The stones are of limestone and it is the chemistry of this rock that holds the key to understanding where shorelines lay in former times. As rain falls, carbon dioxide in the air dissolves to form weak carbonic acid. The rain entered the ancient lake water, making it slightly acidic. Over time this weak acid can dissolve limestone. Limestone boulders at former lake edges were lapped by the slightly acidic lake water, which in time dissolved the bases of the stones. Some of the stones took on the form of mushrooms, although many take other less regular shapes.

4:7 A 'mushroom stone', characteristic of those found throughout the Irish midlands

When lake levels dropped, the mushroom stones marked the larger shoreline of the former lake. Today, 8 km to the north of one lake, a single mushroom stone may mark its former limit.

Around the coasts sea levels were rising. There had been a slow rise in sea water around the coastline of the British Isles since the end of the Ice Age but 9,500 years ago sea levels were still about 30 m lower than they are today. This meant that the Irish Sea was narrower along its entire length and probably at its narrowest somewhere between the coasts of County Antrim and northern England or southern Scotland. It is possible that it was across this narrow sea that people first come to Ireland.

IRELAND'S FIRST PEOPLE

Considering the proximity of Great Britain, it would be foolhardy to state that nobody came to Ireland until about 9,800 years ago (7,800 cal. BC) but there is no record of anybody having settled here until then. A single dilapidated flint flake, similar to those made by people living in southern England, was found in coastal gravels near Drogheda, possibly the only clue to early visitors to Ireland. Even this evidence does not confer on Drogheda the singular honour of being the landing place of the first person to come to Ireland because the flake could have arrived here without any human help. The sea and the movement of glacial gravels could have carried ashore a tool once dropped by someone camping on dry land now covered by the Irish Sea.

It is archaeological studies that trace the first Irish people. Along the northeastern coast, near the remains of ancient camping places, flint tools such as those used in northern England have been found. People may have made the treacherous sea crossing from an English shore, at that time well to the west of the Isle of Man. On a clear day people on that shore could have seen Ireland. It is at Mount Sandel, close to Coleraine in County Londonderry, that evidence left by the first people has been dated, ranging between 7750–7670 cal. BC; however even in the archaeologically favoured north many of the earliest coastal habitation sites must have vanished beneath the rising sea. Along the Shannon and its lakeland stone tools also mark where the first people lived. They may have come into the midlands by boat, moving inland by navigating the rivers.

It has been estimated that the early population of Ireland was never more than 8,000 people, only a fragment of the global population of that time, estimated between 5,000,000 and 10,000,000 people. The early Mesolithic colonists lived semi-nomadically, following the new food brought by the turn of the seasons – but they may not have sought the ancestors of those plants and animals we now think of as edible. Some hints about Mesolithic diet come from isotopic studies of human bones from Ferriter's Cove, County Kerry, which show that people living there ate foods from the sea.

People gathered, hunted and trapped food from woods and open spaces as well as from lake shores, river banks and the coasts, but the pollen record in the lake mud and peat can tell us nothing of the arrival of the first people. Any print on the landscape made by their small numbers was too insignificant to register. In the Midlands the archaeologically priceless rubbish that must have been left by the first hunters and gatherers would have been lost when bogland or river sediments engulfed the places where people once lived.

The first settlers knew how to get the best from the places that provided them with the materials for making tools. They understood the qualities of stone, wood and cordage and were skilled at making serviceable tools that would hold a cutting edge. Larger tools, such as axes, were held in the hand and used to chop into wood. Baskets were used to scoop fish out of water or may have been placed as traps where fish were plentiful. At Clowanstown in County Meath delicately-constructed fishing baskets have been found. They were made by twining twigs of alder, birch and the wood of a species of the rose family, in which the native Irish flora is comparatively rich. The rivers had plentiful salmon, trout and lampreys (*Lampetra* sp.) but the notoriously slippery eels were challenging to catch. Since the times of

4:8 The stones that mark the ancient shoreline of Lough Boora in the midlands, where some of Ireland's earliest people made camp

4:9 Wild pigs disturb soils when searching for nutritlous roots. This photograph was taken in Hungary.

the hunters and gatherers traps made of woven wood strips have been placed in rivers to capture eels and the remains of the oldest fish traps in the British Isles have been found in the estuarine silts of the River Liffey. Salmon and trout were much prized for food and were speared by sharp-eyed people standing on the shallow sandbanks where the Lower Bann River leaves Lough Neagh. At nearby Newferry tens of thousands of stone tools and fish bones have been removed from the diatomite deposits sandwiched within the flood-plain peat. The fish were eaten fresh or possibly smoked, as their oily flesh keeps for months when partially cured. The flesh of wild pig would have contributed high quality protein to the early Irish diet.

The coast was rich in shellfish, crabs, shrimps and small fish. In general shellfish are not rich in calories but they are a good source of vitamins and minerals. There is evidence for shellfish being eaten from shell rubbish heaps or middens, like those at Sutton, County Dublin, and Rockmarshall, County Louth. Sandy and rocky shores were good places to gather oysters, mussels, limpets (*Patella vulgaris*) and dog whelks (*Nucella lapillus*), all fine sources of protein, vitamins and trace elements. Shore crabs (*Carcinus maenas*) and small flatfish could have been caught at the water's edge and larger fish speared in slightly deeper water. From cliff faces in the spring, eggs and chicks of sea birds could be taken. Guillemot (*Uria aagle*) and gannet (*Morus bassanus*) were eaten. Scavenging for food may explain how the remains of a whale ended up in a deposit from a Mesolithic camp site at Curran Point in County Antrim.

The Irish coast is rich in seaweeds such as dulse (*Rhodymenia palmata*) and carrageen moss (*Chondrus crispus*). Amongst the heaps of brown kelp, the bases of the long, frilly fronds of Laminaria saccharina, sometimes called 'sugar foot', could have been chosen as a sweetish food. From shore gravel or salt marshes, the leaves of seakale (*Crambe maritima*), wild sea beet (*Beta maritima*), and the juicy stems of glasswort (*Salicornia* sp.) may also have been collected.

There are about 120 native flowering plants in Ireland that have edible parts. Chickweed (*Stellaria* sp.) is palatable and roots of dandelion (*Taraxacum* sp.) may have served as a food or a medicine. Seeds from wild vetches (*Vicia* sp.) and water lily (*Nymphaea alba*) are nutritious but none seem as valued as hazelnuts, which are a fine source of oil and protein and which can be kept for the lean months of winter as they store well.

4:10
Hazelnuts were a mainstay of the Mesolithic diet

4:11
A modern apple variety is much more fleshy than its wild counterpart

Blackberries may have been eaten fresh as they spoil quickly but firmer fruits like rose hips and wild apples could have lasted into the winter months.

The autumn harvest of mushrooms may have been dried to keep good things in store during the dark days of the cold, wet winters. In spring the protein-rich tissues immediately below tree bark could have provided some nourishment when other foods had become scarce.

There is, however, more to life than food. People need clothes and homes. Hazel provided flexible poles 2–3 m long that could be bent to form the framework of a hut or for fuel for the home fire. Animal skins, when available, must have been very valuable, since a good deal of energy must be expended in their killing and curing. To join materials, cordage was essential. Animal sinews and long fibres from willow bark or bulrush may have been treated to become rot-proof and flexible enough to be twisted into strong twine.

WOODLAND LOSS AND THE FIRST BLANKET PEAT

At the time when woodland was expanding onto bogs in many parts of the country, there were places in Counties Kerry and Donegal in particular, where the hillside woodlands were diminishing. They were giving way to a new type of bogland known as blanket peat, so called because it drapes the land like a blanket. On the western hills the weather was the wettest in the country, the soils had never been fertile and, over the centuries, rainfall further washed the nutrients from the soil. The wet, starved soils were not well suited to trees and as old ones died saplings did not grow to take their place. The woods became less dense and changes within the chemistry and structure of the soils contributed to conditions increasingly suitable for the *Sphagnum* mosses that grew amongst the trees, doing well in damp places and needing only the traces of nutrients that are dissolved in rain.

The woodlands lost vigour and once the *Sphagnum* had taken hold it spread quickly and started to form peat. The heather, crowberry and sedges in the dwindling woodland found an alternative home on the new blanket peat, since these plants were also well suited to life in wet, nutrient-poor, acidic conditions. As time went by the blanket peat spread even into the lowlands. In parts of Counties Mayo and Sligo conditions for the development of blanket peat at sea level were ideal, and below a more recent peaty blanket lie pockets of the oldest of the Lowland Atlantic blanket peat, as this type is called. The large expanses of peat that characterise the modern landscape of Connemara, Sligo and Mayo may have largely developed in later prehistory and the factors that controlled this increase will be considered in a later chapter.

BETTER WEATHER

The period 9,000 to 8,000 years ago had seen a number of changes influenced by variations in climate, but the millennium that followed introduced an improvement in weather that lasted over 2,000 years. Early stratigraphic studies begun in Sweden during the nineteenth century by the geologist de Geere identified a time of renewed warmth in Scandinavia, which he termed the 'climatic optimum'. It was the renowned botanist and naturalist Robert Lloyd Praeger who realised that Ireland had also enjoyed the same warmer conditions. His findings were based on layers of snail shells that turned up in estuarine mud in Belfast Lough when the Alexandra Dock was being built in 1887. The excavations revealed a bed of peat 8 m below modern sea level, providing proof of earlier low sea stand along the Belfast coastline. As the excavations went deep into the estuarine clays, layers of snail shells pointed to very early times when the waters of Belfast Lough were especially cold and also to a period starting about 7,500 years ago, when the Holocene climate appears better than it had been before or since. Snails are most exacting in their temperature requirements and they die where they lived. If the shells of snails that live only in warm conditions are found trapped in deposits in places that are now cold, one may assert that, at the time when the snails were alive, the weather was warmer. Praeger's studies confirmed that early change in climate. During the 'climatic optimum', the woodlands were at their most luxuriant. Oaks, elms and pines grew where soil suited them and numerous, smaller tree species grew beneath the woodland canopy or on the forest edge.

THE RISE IN ALDER

Up until this time alder had been a rare tree growing near water but its status was about to change. Pollen records from the Irish lowlands, dated to about 7,000 years ago, show increasing amounts of alder pollen preserved in peat and lake mud. These pollen records show that something was letting alder rise from a position of little importance to one in which it became the most widespread tree able to cope with extreme wet. In Ireland there is only one native species of alder, *Alnus glutinosa*. It makes a large tree with distinctive round, shiny leaves that have notches rather than points at their tips. Alder is a much more robust-looking tree than the willow, previously the most common tree species growing close to water. Throughout the British Isles the expansion of alder marks a major woodland revision and the rise in the fortunes of the tree coincided with a shift to a damper climate.

Alder is a tree that favours waterlogged soils on the banks of lakes and rivers. Oxygen-depleted, saturated soils are low in nitrates, a nutrient vital to tree growth, but alder can live under these conditions because it has a relationship with an

4:12 *Nodules on the roots of alder that contain the actinomycete*

actinomycete (*Frankia*), a primitive micro-organism similar to a fungus that lets the tree meet its own nitrate requirements.

On alder's roots are woody growths or nodules about the size of a walnut, within which live the actinomycetes that trap the nitrogen in the air and convert it into plant food. Having an inbuilt source of nitrogenous plant food provides the tree with a survival strategy, well suited to life in wet soils where nitrates may be at a premium. Additionally alder's presence will return fertility to the soil as its leaves decay each autumn.

At the time when alder expanded its range, it may have done so because soils in lowland valleys and estuaries became wetter as the climate got damper. There is, however, another explanation. Throughout Irish prehistory there is evidence that variation in lake levels was loosely linked to climate change. Falls in lake water levels would have exposed fresh, wet ground, providing ideal conditions for the germination and growth of alder seeds. Dating of the expansion in alder shows that its spread in time was patchy and it may have taken two thousand years or more for alder to reach upland areas.

FURTHER CHANGES AT THE COAST

Changes in levels of water were not restricted to lakes or rivers. Throughout the period 9,500 to 6,000 years ago the shape of the coasts also varied as sea and land levels rose and fell. Sea level change is imperceptibly slow and controlled by processes involving rises in the volume of water in the sea, expansion of the sea's upper layers as water warms, and uplift of the land when pressures from dwindling ice sheets are reduced. Experts differ in their estimations of where sea and land levels were at any chosen point in past time, therefore generalisation is necessary. There are raised beaches along the northern coastline that mark where the land was once higher by many metres. The best and highest Holocene raised beaches are

4:13 At the Giant's Causeway, the upper limit of the raised beach shows as a line marked by change in vegetation

said to be around the northeast Antrim coast but they have sometimes been confused with raised platforms in the ancient lava in which the Antrim coast is rich. At the Giant's Causeway the raised platforms are reasonably evident and fall approximately on the projected mid-Holocene bench line. And the platforms are often defined by perched freshwater, marked by a distinctive change in vegetation.

The signs in the north are more pronounced than on southern shorelines and hint that the land's uprising may have happened in jerks. With a few exceptions around Bantry and Galway Bay, however, post-glacial raised beaches are rarely found on the west coast.

Between 8,000 and 7,000 years ago sea levels may have been more or less where they are today but by 6,000 years ago, sea levels were higher than at present. This 'marine transgression', as it is called, happened when the sea was rising more rapidly than the land, swamping parts of the coast perhaps under three metres of water. The sea carried materials such as shells to places now inland and Icelandic volcanic pumice that had floated across the North Atlantic has been found on a raised beach at Lough Swilly. Around the coast sea water swept into estuaries, pushing river water back along the river channels, raising the level of the river waters. Flooding of the surrounding land formed tidal lagoons, moved sand bars and smothered fens and their underlying peat under mud or gravels.

4:14 Fossilised pine stumps embedded in the beach at Spiddal, County Galway, mark a place on the coastline where sea levels have varied in the past

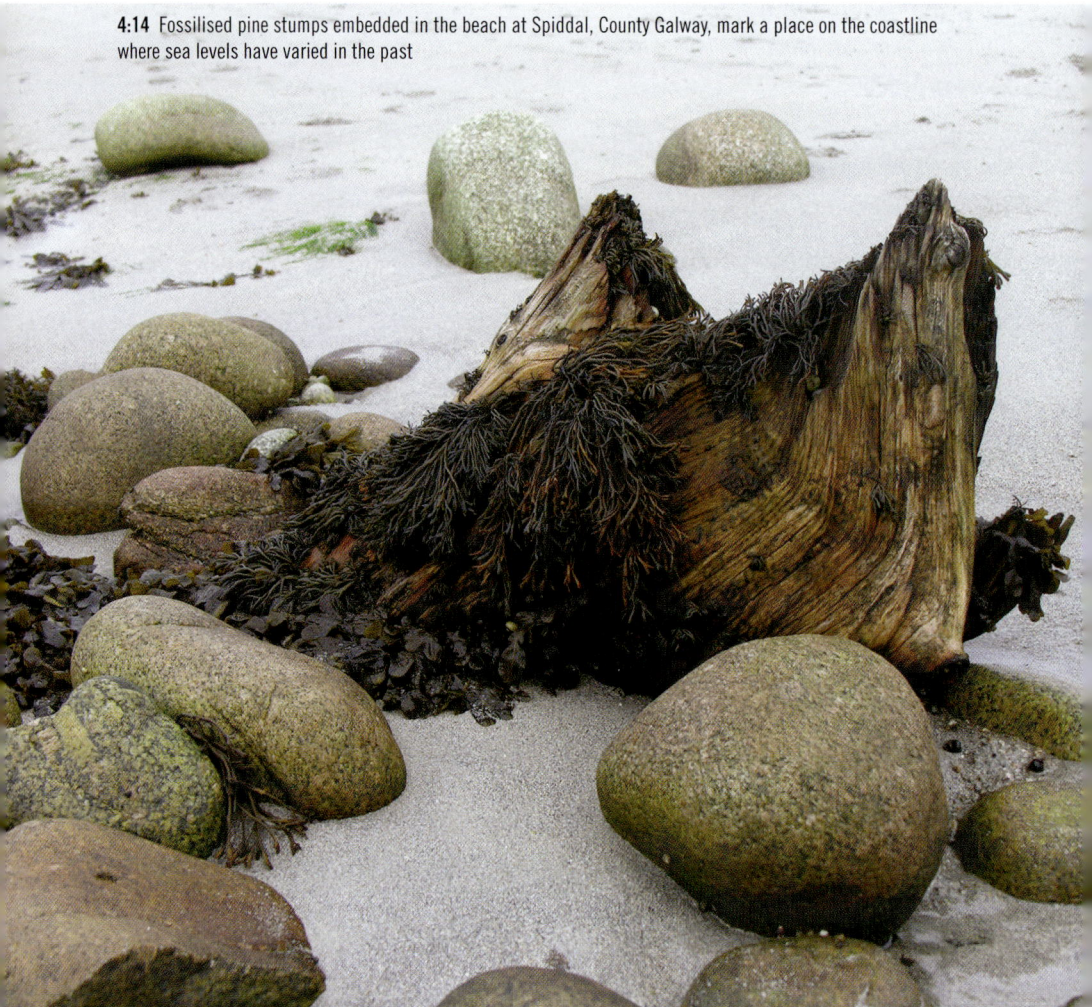

FURTHER CHANGES IN THE UPLANDS

By about 6,000 years ago there had been further changes in the uplands in particular. The blanket peat had spread at the expense of the upland woods and, by then, the first signs in the pollen record appear to show that people were beginning to alter the landscape for their own purposes. One of the few ways of distinguishing natural woodland openings from those made by people is to look for signs of burning.

Pine and hazel grew in many upland places and both burn well, with hazel re-sprouting readily after burning, growing to maturity and producing valuable crops of nuts. Where increased signs of clearance regularly included findings of charcoal, the evidence pointed to people manipulating the landscape, albeit on a modest scale. They appear to have changed their lifestyles so that they lived more or less permanently in lowland camps, from where they journeyed afar and into the hills.

4:15 A layer of black charcoal at the bottom of a blanket peat deposit

There are various reasons why people may have made openings. By 6,000 years ago it may be possible that there were sparse herds of red deer in the uplands, although the evidence for the presence of this animal in late Mesolithic times remains contentious. If the animals were present, the openings may have attracted them to good grazing and the possibility of a kill could have been improved. Or it may have been that choice plants that grew best in the open were encouraged by clearing a bit of land. In the west of the country there were also signs of more frequent fires, hot enough to leave black scorch marks on pine stumps. The fires could have been natural but taken with the other evidence for a landscape undergoing change it is probable that some were started by people. These clearings are of great significance as they mark the start of a new relationship between the environment and its people. In the centuries that followed, changes wrought by people would change the character of the landscape completely.

CHAPTER 5

THE INFLUENCE OF EARLY FARMING

6,000–4,000 years ago

KEY ISSUES:

early farming and the first major loss of woodland;

the Elm Decline;

early farming;

Céide Fields.

TIME PERIOD FOR
THIS CHAPTER

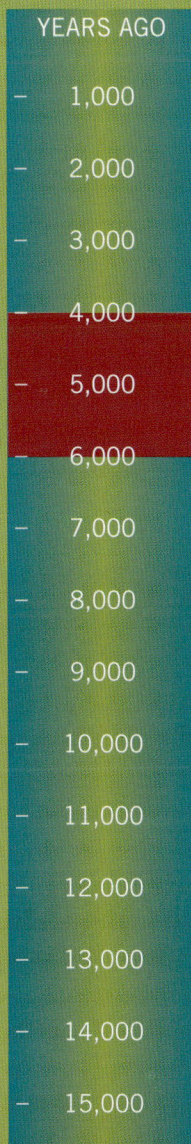

YEARS AGO
1,000
2,000
3,000
4,000
5,000
6,000
7,000
8,000
9,000
10,000
11,000
12,000
13,000
14,000
15,000

FARMING BEGINS

Thse first attempts at landscape management described in the preceding chapter were insignificant in their impact when compared with the changes that happened once farming was underway. The loss of woodlands and the subsequent creation of open spaces in which crops were grown or animals grazed were the first indications of changes to the Irish landscape that continue to this day.

Farming was brought to Ireland less than 6,000 years ago, probably from Great Britain or Europe. The type of farming practised in Europe began in the Middle East not long after the Ice Age ended, when the world's climate warmed rapidly. It was first recorded in the area known as the Fertile Crescent, which starts in the foothills of the Zagro Mountains in southwestern Iran, moves in a great arc northwards through southern Turkey and ends in the valley of the River Jordan. Nine thousand years ago the people of this area started to domesticate animals and plants, for example cattle, sheep, goats, pigs, wheat and barley. Farming is harder work than hunting and gathering but the compensation for the effort expended comes as more food. The needs of domesticated animals and plants, however, placed new pressures on the landscape's resources. It behoves us to remember that almost half of the 11,500-year duration of the present interglacial had elapsed before human pressure began to alter the Irish landscape.

In the archaeological record the start of farming is one of the markers for the New Stone Age or Neolithic period. Farming started with the introduction of animals and plants that had already been domesticated. Cattle and cereals were brought here by people, but by whom or from where is far from clear. The first cattle (*Bos taurus*) bones found anywhere in the British Isles come from Ferriter's Cove on the Dingle Peninsula in County Kerry where cattle leg bones have been found, radiocarbon dated to over 6,200 years ago (4500–4232 cal. BC). This first bone evidence for cattle in Ireland may not herald farming, as a cow could have been a status symbol or the bones could have originated from imported meat.

THE NEW AGRICULTURAL LANDSCAPE

The first agricultural land may have been carved out of the woodland by increasing the size of the natural clearings. The polished stone axes used by the Neolithic people were quite well suited to cutting into slender tree trunks. The broad axe edges were keen but the cutting surface was at a shallow angle to the blade and not ideal for biting deeply into thick tree trunks. Felling a tree rarely kills it because dormant buds in the wood of the stump sprout to produce leafy twigs or the roots throw up suckers. If, however, a bit of opened woodland was grassy enough to feed a few head of cattle, their browsing would soon get rid of any young tree growth, thereby keeping the cleared area open.

5:1 Johannes Iversen, noted past environmental scientist, using a polished stone axe to cut down a tree. The photograph was taken in the 1950s at Reading University.

Where big trees grew, the land was harder to clear but it was vital that the large trees were killed because in woodland the bulk of the plant nutrients are locked up in the tissues of the trees. One way that the pioneer farmers could have released these precious plant foods to the soil quickly was by burning dry, dead timber. A slower but effective method of killing a large tree is by starving it to death, achieved by cutting deeply through the bark of the tree and removing a ring of living tissue. The practice is called ring-barking or girdling and it works through disrupting the tree's food supply by severing the phloem, the tube system beneath the bark that carries the food round the tree. A weakened oak or elm would have been further hastened to its death as fungal infections invaded and weakened the wood and every storm ripped branches from the dying tops.

The conditions of those long-past summers were ideal for the fires that reduced the timber to a carpet of wood ash, which then fed the young grass foliage so nutritious for the cattle. Burning was detrimental to the soil in the long-term, as vital nitrates were lost to the atmosphere and potassium and phosphates in the wood ash were washed into the soil by rain. The organic content at the soil surface also suffered as fine particles of charcoal blocked the tiny air spaces, reducing the soil's porosity. Through fire, forest soils that had formed over the previous 4–5,000 years could be changed within days.

People worked to prepare the land for crop and stock farming but how they did so is poorly understood. The winning of agricultural land from the wild is termed 'landnam', a word taken from the old Danish language. The Neolithic landnam did not happen simultaneously across the country. In Irish pollen diagrams 'landnam' may be marked by falls in levels of tree pollen, sometimes accompanied by slight rises in grass pollen values. As well as these changes there were rare findings of pollen grains that may or may not, for this is most contentious, have come from primitive wheat. Overall, the archaeological and pollen analytical evidence underscores that the earliest farming was on a small scale.

Once the soils became depleted, the farmers abandoned the land for some years and the growth of shrubs and trees began to refresh the exhausted soil. The roots of these plants penetrated to the deeper, richer soils and drew nutrients up into their leafy tops. In autumn nutrients were returned to the soil surface in the decaying leaves. Each year's leaf fall enriched the top soil and the cycle of agriculture started again as the scrub was later burnt in preparation for pasture or tillage.

THE ARRIVAL OF THE CATTLE

It must have been in the early part of the fourth millennium BC that someone took the bold decision to bring live cattle into the country and possibly by boat. Transporting a live calf in a small boat, even when the beast is tied to stop it struggling, is a hazardous exercise but perhaps the pioneering farmers knew enough about their destination to have faith that their efforts would be worthwhile. Whether the cattle were kept only for their meat or also perhaps for their milk remains a contentious issue in early Irish archaeological studies. The interaction between the need to provide grazing and the impact on the wooded landscape will be the subject of a later section in this chapter.

Facing page:
5:2 Ring-barking will weaken and ultimately kill this tree

THE FIRST CEREALS

In Irish pollen diagrams it is an increase in the percentages of native grass pollen that hints at new pasture but the pollen record for the earliest wheat has proven difficult to interpret. Wheat originated from the wild grasses of the Fertile Crescent and although its pollen grain is the same shape as a grass, it is much larger than that of its wild relatives. The largest pollen grain produced by a wild grass is almost the same size as the smallest pollen grain produced by a cereal and therein lies the difficulty. How does one confidently separate large wild grass pollen grains from the precious few that came from the first wheat crops?

Efforts to resolve the problem by close study of the pollen grains in Irish peat and lake mud seem to have made the difficulty worse, for the more the oldest sediments have been studied, the more they show that they contain cereal-sized pollen grains. Where, however, the disputed pollen grains turn up in the earliest Neolithic pollen profiles, pollen grain size and shape, microscopic characteristics and the other plants represented in the pollen profile have all been used to evaluate claims to cereal status and there is dated evidence that the first cereals grown in Great Britain and Ireland could have been between 4800–4500 cal. BC.

Growing cereals is hard work, as the crop needs to be sown in well-broken soil. Early farmers had the arduous task of opening root-infested ground using wooden digging sticks, for the simplest ploughs had not yet entered Ireland. The digging

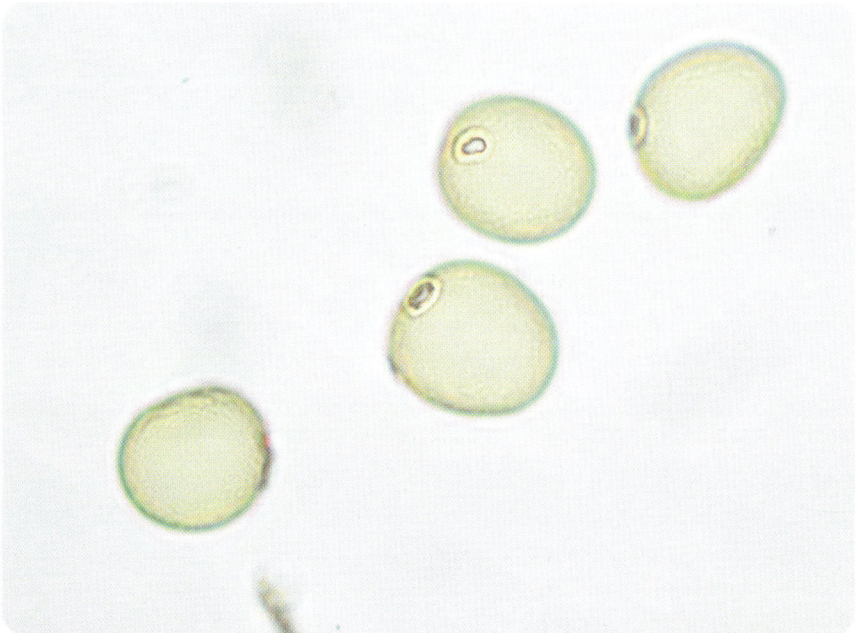

5:3 Light micrograph of modern wheat pollen grains

5:4 The house mouse has lived in Ireland since the time of the first farmers

sticks were probably about the length of an adult's forearm and sharpened to a point hardened in the fire. They were used to scuffle patches of soil to prepare them for seed sowing. The seed may have been drilled into the ground, because wheat seed was not plentiful. Drilling rather than scattering also gives a better return at harvest, when the ears all ripen at the same time.

Primitive cereals were remarkably resilient grasses and, in Ireland, emmer (*Triticum dicoccoides*) and barley (*Hordeum* sp.) as well as einkorn (*Triticum monococcum*) and bread wheat (*Triticum aestivum*) were probably grown. The little cereal seeds germinated quickly. Clumps of closely-spaced cereal plants would seem an ideal home for pests but primitive wheat is almost immune to attack by fungal rusts or aphids.

Unlike wild grasses whose seed-heads fall apart when ripe, early wheat's seed-heads stayed unbroken, making it easier to harvest the intact ears. The crop may have been gathered by snapping the ear from the stem or cutting off the seed-head with a sickle made from a stick edged with small, sharp stone flakes. Once dried, the seeds lasted for months without spoilage by fungi or insect pests, although some of each season's crop must have been devoured by the house mice (*Mus domesticus*) that came into Ireland with the first farmers.

Every stage of cereal cultivation was laborious, from soil preparation to grinding the seeds between stones to smash the outer coat and release the nourishing starch and protein in the seed embryo or 'wheat-germ'. The gritty flour was then ready for cooking, perhaps baked on a hot stone into flat bread or simmered as porridge, possibly with wild vegetables, fruits or meat.

THE FIRST WEEDS

Weeds are the abiding companions of cereal crops and although the majority of species that became weeds were Irish natives, new weeds may also have arrived with the first imported cereal seeds. Nevertheless, the first crops may have been reasonably free of weeds as it may have taken weed seeds years to build up in the soil.

5:5 Neolithic fields may have been almost as free of weeds at those of today, when treated with weed-suppressant chemicals

Stores of cereal seeds found at Neolithic sites are almost free of weed seeds, so it seems that poppy flowers (*Papaver* sp.) may never have speckled the first ripening crops as they did in historic times.

In possible contrast to poppy, native ribwort plantain (*Plantago lanceolata*) is the one weed that did not need time to become established. The pollen records for the earliest farming periods throughout Europe show an upsurge in ribwort plantain pollen. Ribwort plantain has grown in the British Isles since the end of the last Ice Age and is well suited to growing in disturbed soils close to glaciers, conditions mimicked by the soils of the first farms. The plant has a flattish rosette of leaves that can withstand damage by hooves, for bruised leaves are soon replaced by others, letting it grow where other plants would have failed. Field observation shows, however, that the leaf length and uprightness of habit varies depending on associated vegetation.

It is possible that the 'seed corn' may have been an additional source of plantain seed. In Ireland it is hard to tell if particular weeds were unique to cattle-keeping or cereal cultivation, since today the ribwort plantain grows on the edges of crops as well as amongst the meadow grasses. Throughout Europe ribwort plantain's characteristic pollen grain is seen in pollen analytical studies of deposits that accumulated when farming was first practised.

Left:
5:6 The characteristic rosette of the leaves of ribwort plantain

Below:
5:7 When surrounded by grasses, the ribwort plantain's leaves assume an upright habit

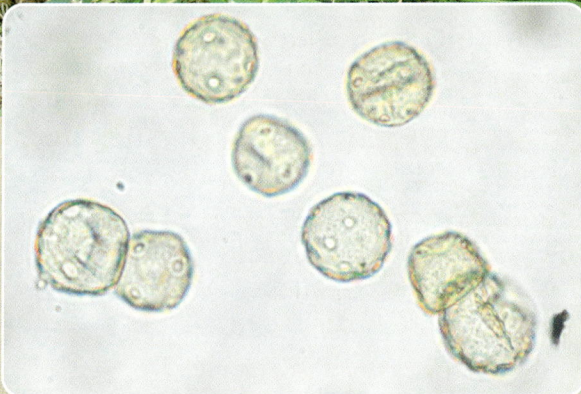

Left:
5:8 A light micrograph of the pollen grain of ribwort plantain. The pores in the pollen grain wall are characteristic of this species of plantain.

Farmers understand the importance of good soil, heightening the expectation that early people preferred to live in fertile areas. It is easier to spot where people once lived through the location of their great tombs that still dot the landscape. In County Leitrim, the tombs are on the best soils but in parts of west Ulster many are on the less fertile land, indicating that soil type was not the only factor that influenced where people lived and farmed. There is also the possibility that cereals were grown further upslope than would be possible today. Cereal-type pollen dated to 3647–3156 cal. BC was found near the bottom of a blanket peat, 485 metres up Slieve Croob, in mid-County Down. Not that all mountain slopes were used for farming as many were still under scrubby woodland. At the same time as cereals may have been grown on Slieve Croob, the slopes of Slieve Gullion in County Armagh were covered in hazel scrub.

It was stated previously that the early stages of farming are not fully understood but the heated arguments that raged in archaeological and palaeoenvironmental circles during the 1960s cooled as new findings emerged. Better-dated evidence makes plain that there is no single agricultural model that applied to the entire country. In some places cereals were grown before cattle farming was introduced, whilst in others the opposite applies. It is now believed that many of the first farmers lived settled rather than nomadic lives. The latest research is investigating the possibility that people were farming within the woodlands rather than getting rid of them to open spaces and soils for agriculture. There are persuasive cases from Europe for what has become known as 'forest farming' and we await with interest the latest findings that may show if that system was used in Ireland.

Although the parts of the landscape that had been opened changed appearance because of little strips and patches of ripening grain, throughout the country there were large tracts of terrain that remained free of agricultural pressures. On the heavy soils of the densely wooded lowlands nobody worked the heavy, poorly drained land. The lake edges kept their wet woods and the trees and grasslands on the upper wind-swept mountain slopes also remained unaltered. Much of the countryside was left only to the birds and the insects.

THE DECLINE IN ELM TREES

At the time when the extent of open land was expanding, something strange happened to one of the trees in the woodland. Pollen diagrams that record vegetation changes at 3840 cal. BC throughout the British Isles and nearby Europe show a drop in the levels of elm pollen – the event known as the Elm Decline. A hint that the downturn in the fortunes of elm might be linked to the new farming practices came from the tree's preference for good soil. Places where elm did well may also have been attractive to people grazing cattle.

One explanation for the fall in elm percentages was that people had cut the elm's young branches to feed their cattle. As a result of repeated cutting, elm trees sprouted new wood that never had the chance to flower. But there was a snag with this explanation. To account for such marked falls in elm pollen percentages, it would have needed people with stone axes lopping the flowering wood away from almost every elm tree in Europe. The decline of the elms was very widespread and the fact that it seems to have happened almost simultaneously in all areas might suggest a climatic cause, such as late spring frosts killing the young flower buds. A further hint of a deteriorating climate comes from the oaks that had been growing increasingly abundantly on bogs at that time, as their numbers also went into decline.

An alternative explanation was that the trees had been devastated by the ravages of a fungal infection, possibly Dutch elm disease, which killed thousands of elm trees throughout the British Isles in the 1970s. The disease is caused by a type of primitive parasitic fungus (*Ophiostoma ulmi*), which may be carried by two species of bark-boring beetles (*Scolytus scolytus* and *S. multistriatus*), both of which are strong fliers.

5:9 *Scolytus scolytus*

Once the beetle has traced the tree by its smell, it bores holes in the elm tree's bark and lays eggs that hatch into grubs that devour the soft wood beneath, leaving a network of tunnels and galleries. The tunnels allow the grubs' lethal loads of parasitic fungal strands and spores to be carried through the large tubular vessels in the spring-grown wood of the tree.

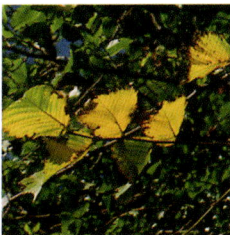

5:10 The yellow leaves of this elm branch show that it is infected by Dutch elm disease

If the disease stays in the branches death can be slow, but if it reaches the roots and gets into the sap system the afflicted tree dies faster because small yeast-like cells of the fungus travel up the tree in its sap. It is thought that elms have a defense system that prevents micro-organisms reaching and colonising actively growing parts but the fungus can either suppress or overcome the defense mechanism. The first sign that a tree has sickened shows in its leaves, which wilt and turn yellow at the height of summer. Once the infection gets a grip, the tree is doomed and even the largest will be dead within a few years.

If the Neolithic elm decline was caused by Dutch elm disease, its spread and deadly impact may have been aided by people harvesting the foliage, since the vigorous young growth from cut wood is very susceptible to infection.

In the west of Ireland the elm decline is sometimes separated from any sign of human impact on the landscape, thereby underlining the evidence for a disease cause. No conclusive evidence older than 5,000 years has been found linking the beetle, fungus and elm timber in Ireland, but wood of that age with elm bark beetle galleries has been identified from Denmark. In Ireland the circumstantial evidence for a disease cause for the elm decline is persuasive. The loss of many mature trees then and again a few hundred years later must have had a profound effect on the insect fauna of the woodlands and especially on those that depended on the oldest trees. By 5,200 years ago, when the phase of intensifying farming had come to an end, agriculture went into a lull. Elm had recovered in many woodlands, possibly because new disease-resistant strains had emerged, as may have been the case in the woodlands near Littleton Bog in County Tipperary and Scragh Bog in County Westmeath. In the north elm regained ground in some places but not to its former extent. The soils that had suited elm well were depleted of nutrients by agriculture and where these conditions occurred, oak and hazel grew back instead.

New evidence from research in Scandinavia, on bones from early cattle and aurochsen, sheds further light on this debate as there was nothing in the isotopic analysis of the cattle bones to show that they had been fed on elm. The isotopic evidence indicates that the earliest cattle had been fed on grass but comparable information is not available for Ireland. The aurochs is not a native mammal and there is a scarcity of cattle bones in the archaeological faunal record, confounding attempts to reconstruct early stock farming.

CATTLE FARMING IN COUNTY MAYO

The elm decline marked the start of an intensive phase in early Irish farming, with much 'landnam' activity, which in parts of the west may have lasted for nearly 500 years. It was during this period of intense agricultural activity that there emerged one of the most extraordinary examples of Neolithic farming anywhere in Europe

5:11 One of many Neolithic field walls at Céide exposed when the overlying blanket peat was removed

and all in the interests of keeping cattle. The stone-walled field system at Céide represents the oldest enclosed farmland in the Western world. The walls were discovered during the early 1980s, at Céide near Belderrig, above the cliffs of the coast of north County Mayo. Before the discovery of the walls and fields, there was nothing in particular that hinted at their presence, for they were smothered by blanket peat indistinguishable from the surrounding bogland. Extensive probing of the peat with rods has proven the field system to be enormous, covering more than 1,000 hectares, with the entire system having lain undiscovered beneath the bog for over 5,000 years. The fields are large at Céide, between 5 and 50 hectares, and all are bounded by a grid of walls made from low mounds of soil and rubble, topped by flat stones. To the west there are other places where stone walls and earthen banks define yet more of the first farmers' ancient fields.

Before the fields were made, places at the back of the cliffs were already boggy but most of the area was lightly wooded, with pines and scattered shrubs growing on the poor mineral soil. It is worthy of note that the fields at Céide were enclosed by countryside in which clearance by people contrasted with untouched woodland. The trees and shrubs had to be cleared but not by the use of fire it seems, for there is little evidence of burning at Céide. The regularity of such an extensive system of fields must mean that it was constructed to an overall plan. People put great effort into removing the woodland and they carried massive quantities of earth and stones that both cleared the fields and provided the material to make the walls, most of which were built on the mineral soil. The field system may have been laid out within the lifespan of one or two generations and its extent indicates that numerous people committed themselves to this extraordinary enterprise.

As the years passed, the fields stretched further and the growing herds of cattle must have been grazed all year round. It is very likely that the cattle were farmed for their meat but it is not impossible that occasional milking took place. The pollen record shows that fields were weedy with clover (*Trifolium* sp.), buttercups, dock and the numerous types of plants that look like dandelions growing amongst the grass, meaning that the first big fields in Ireland had an appearance similar to modern rough pasture.

At its height at Céide, farm practice had diversified into a few small fields of wheat. The farmers had progressed from digging sticks to using simple ploughs, which have left their marks in some of the fields. The simple plough or 'ard' can be considered a type of mechanised digging stick because it scratches grooves into the soil and prepares it for seed sowing. At Céide a bit of broken stone from the cutting tip of an ard has been found, although it is not know if this was in use at the time when the fields were ploughed. After five hundred years or so Céide's great cattle farm had dwindled away and there is no clear explanation for its demise. The soils were never fertile or fertilised with imported material and there is nothing that points to bad weather pushing the system into decline. Whatever the reasons, it seems that a failing landscape was not one of them.

Bit by bit the blanket bog began to grow over the ungrazed field. As the years passed, the peat deepened until the walls and fields were seen no more. Fields that had once been white with clover flowers became stretches of blanket bog, then white with the blossoms of the bog cotton. In some places the peat grew quite swiftly, whilst in others its progress was slower. It seems puzzling that the peat grew so well when the weather was generally dry but the loss of pines when the fields were made must have favoured the bog mosses. Before the pines were stripped away, they mopped up the rainwater, which evaporated into the air through their leaves. After the trees went, the rainwater remained and better served the growth of *Sphagnum*.

Although dilapidated field walls may be seen disappearing beneath blanket peat in many parts of the uplands, there is nothing like the Céide field system anywhere else in the country, meaning that the practice of cattle-keeping in the west was probably unique. An alternative to keeping cattle in the same fields all year round was to take them to lush upland grassland over the summer months. This practice has the advantage of making use of good summer grazing and keeping marauding beasts off the precious wheat. On the slopes of the Antrim Plateau it is thought that cattle were moved upland to crop the summer pastures and then brought back to the dwelling area as the growing season ended.

THE AFTERMATH OF THE DECLINE IN FARMING

At this time at Lough Neagh and in north County Antrim, increasing numbers of oaks grew on the fens and the edges of the big bogs. The bog oaks may not in all instances have been as large as the trees on the dry land because they were poorly nourished. Where bog oaks grew in loose groups they were accompanied by the occasional willow, whilst further out on the bog pine trees continued to grow. Amongst the living pines there were dead trunks still standing upright and others that had fallen as the bottoms of the trunks had rotted through, as at Sluggan Bog in County Antrim where pines continued to grow on the bog surface between 5,300 and 5,200 years ago.

Yews and, to a lesser extent, ash did better in the south and west than anywhere else in the country. Yew expanded into woodland where it had been a rarity, for example in Reenadinna wood in County Kerry the invading yew trees did exceptionally well. This yew wood has kept its ancient character and is today one of only three yew woods in modern Europe. From Muckross in County Kerry to the Burren in County Clare, yews sprang up and, although they started to peter out as far north as County Roscommon, here too they may have formed small patches of woodland. They did not take over bog surfaces as pines had done from time to time but big yew trees most definitely grew in places in the big midland bogs.

Along the west coast the broadleaved trees did not return in abundance to their old ground. On the Dingle Peninsula, the land went over to heathland, bog and pasture. In the wettest parts of the west and north, blanket peat gained ground and on some of the slopes the new peat started to erode not long after it had formed.

Facing page **5:12** The trunk of a substantial bog yew that had once grown on a midland bog

In parts of County Longford the wet lowland bogs had become so treacherous that people took action to make travelling across one of them a little safer, through building wooden walkways, of which more will be described in a later chapter. Cut marks on the walkway timber showed that they used stone axes to slash through hazel, alder and birch stems and brushwood from the woods near Corlea Bog. The wood was tightly packed into layers to make these first trackways. In two places on the bog surface the brushwood was laid down in a strip 1.6 m wide, supported by occasional strong timbers, with the edges held down by a few wooden pegs.

As the centuries passed and the Neolithic period drew to its end, woodlands near places where people were engaged in big building projects must have come under great pressure. At Newgrange on the bend of the River Boyne in County Meath, the building of the elaborate tombs must have needed timber drawn from the surrounding woods. The land close to the tombs had been previously under grass and there must have been farmed land somewhere in the area, otherwise where would the meat and flour needed to feed the tomb builders have come from?

The story of the impact of the early agriculture on the Irish landscape draws greatly on inland sites but the coasts were important places to people at that time and these places too were changing. In the early Neolithic, during her thirty-year life-span, a woman living inland would have experienced the changes around her home brought about by agriculture. Had she lived by the sea, however, she would have been unaware that the changes to the coast, going on since the retreat of the glacial ice, were drawing to completion. The fall in sea level in the north of the country, which began 6,000 years ago, helped the onshore movement of coastal sands and played a part in the growth of the Irish sand-dune systems. Between 6,000 and 5,000 years ago the outline of the Ireland with which we are today familiar was more or less in place, although by the end of this time the sea had briefly fallen back to a level somewhat lower than today.

The long-established, wooded landscape that still covered much of the country even when farming was at its most extensive was soon to slip further into the past as the landscape became subjugated to the needs of people. The weather was about to worsen and pressures on the woodlands were about to increase in order to provide the fuel to make bronze and gold, the harbingers of the first of the ages of metal.

CHAPTER 6

THE LANDSCAPE OF THE IRISH BRONZE AGE

4,000–2,700 years ago

KEY ISSUES:

woodland loss and regeneration;

the decline of pine;

spread of blanket peat;

developments in farming.

TIME PERIOD FOR THIS CHAPTER

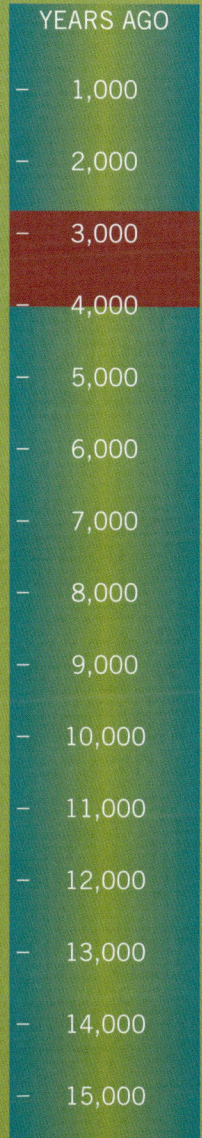

YEARS AGO

- 1,000
- 2,000
- 3,000
- 4,000
- 5,000
- 6,000
- 7,000
- 8,000
- 9,000
- 10,000
- 11,000
- 12,000
- 13,000
- 14,000
- 15,000

Beneath the blanket peat in the uplands may be found chunks of wood and layers of charcoal. The fossilised wood proves that trees once lived in places now clothed in heathers and bog-mosses and the charcoal layers show that fire played a part in the vegetation change. Four thousand years ago processes were underway that saw the upland woods replaced by blanket peat. At this time also the upland pine woodlands were declining, more land was being brought under cultivation and the climate was deteriorating.

FURTHER WOODLAND LOSS

It seems that woodland depletion was further exacerbated by agricultural expansion and worsening climate. Trees were cut and burnt to open new farm land but these activities damaged the soil as drainage became impeded, in part because the air spaces between the soil's crumbs became blocked by microscopic particles of charcoal. Eventually the soil became exhausted as it lost nutrients to grazing, tillage, and leaching by the rain. Tracts of depleted land were abandoned and fresh areas opened by yet more felling and burning, to let farming continue. With the passing years the badly-drained, oxygen-depleted soils became increasingly unsuitable for tree seed germination or sapling growth, explaining in part why, in general, trees did not recolonise abandoned farm land as readily as they had done in the past.

THE EFFECTS OF MINING ON THE LANDSCAPE

The impact on the woodlands of felling by the metal axes coming into usage is hard to judge but could the manufacture of bronze have contributed to the reduction in woodland? To make bronze, one needs fuel. Bronze is an amalgam of about 80 per cent copper and 10 per cent tin, which had to be obtained from Cornwall, one of the few places in Europe where it is plentiful.

The first Irish metal tools were axes made from native copper from the Cork and Kerry region. Copper ore is fairly common in deposits throughout Ireland, for example in the County Tyrone uplands, but there is nothing to show if it had been mined there in prehistoric times. The earliest copper mines in the British Isles are at Ross Island on the Muckross Peninsula in County Kerry and operated from 1700–1500 BC. At Ross Island, wood was the fuel used in the metal-making processes as heat was needed to split the ore from its parent rock as well as to smelt the metal from the ore. To fuel the fires needed for both processes, it has been estimated that about 7.5–10 tonnes of wood were consumed annually at the Ross Island site, presumably cut from nearby woodland.

6:1 An Irish Neolithic stone axe head and an Irish Bronze Age axe head. The blade of the metal axe has a more slender shape.

In the first stage of mining fires of willow or hazel wood were built against the ore-bearing faces so that the heat could split off slabs of rock. The miners used stone hammers to pound the heat-fractured ore and the large scale of this part of the mining process may be gauged by the thousands of stone cobble hammers littering the sites. Cattle shoulder bones served as shovels to lift the broken rock into leather or wooden containers, which were then carried to the smelting fires.

6:2 A cattle shoulder bone such as this would have served as a miner's shovel

Almost pure copper metal was smelted from the ore using charcoal – a further indication of the need for wood at all stages of metal production. Bronze tools were formed by pouring the liquid metal into moulds carved from stone. Like the stone axes of earlier times, a metal axe head had to be mounted and secured to a wooden handle, using cordage. Long, slender copper axes heads were well suited to felling trees but they did not hold their sharp edge well.

The landscape close to the Kerry mines appears to have been fairly open. The fuel may have come from managed woodland, since tree-ring studies of timber from the site suggest coppicing. Charcoal used in the smelting processes may also point to woodland management nearby, although charcoal has the disadvantage of crumbling easily and not transporting well. Charred splints of resinous pine wood have been found at the mines, suggesting that they were burned to give light. Throughout the 200 year period when the Ross Island mine was open, between 1,500 and 2,000 tonnes of wood may have been used. There is, however, nothing in the local pollen record to show that mining needs were destroying the woods. It seems therefore, that mining, whether for copper or gold ore, played no great part in the woodland loss that occurred in Ireland at that time.

WOODLAND REGENERATION

Where woodland grew back again after felling, the mix of species and age range of the trees that made up the woodland had altered and such changes must have had an influence on bird and insect species. The fossil insect record shows fewer of the species dependent on very mature trees – those between two and three hundred years old. Hardest hit were the wingless insects and those that lived on the forest floor, where the loss of leaf and twig litter destroyed the insulation needed by insects that hibernated. Clearances opened the woodland to greater extremes of temperature and moisture loss, which in turn may have altered the frequency and intensity of forest fires.

The fossil bird bone record for prehistoric Ireland varies from place to place with most coming from archaeologically investigated sites. The rich assemblage from the excavation at coastal Dun Aonghasa yielded bones of guillemot (*Uria aagle*), along with razorbill (*Alca torda*), puffin (*Fratercula arctica*), shag (*Phalacrocorax aristotelis*), cormorant (*Phalacrocorax carbo*), kittiwake (*Rissa tridactyla*) and herring gull (*Larus argentatus*).

From time to time the bird bone record reveals surprises. A radiocarbon-dated bone of a great spotted woodpecker placed the bird in the Bronze Age. Woodpeckers live amongst trees and this tiny bone came from County Clare, a county noted in modern times for having areas spectacularly devoid of trees. From this bone record we have a hint that birds no longer native may once have been so.

6:3 Bird bones in a similar condition to these were found in a Bronze Age archaeological context

As a further consequence of the loss of the woodland habitat there must have been a loss of bird and insect species, such as wood ants, that could not adapt to other environments. On the other hand some bird and mammal species probably benefited from the newly open environment. Populations of ground-nesting birds like the skylark (*Alauda arvensis*) may have increased as pasture and bogland extended and hares would also have benefited from a more open landscape. Cereals may have given a home to the corncrake (*Crex crex*), mice would have done well where grain was spilt and the increase in open pasture suited the rook (*Corvus frugilegus*).

THE PINE DECLINE

At about the same time as the upland woods were dwindling, the Irish landscape almost lost one of its longest-established native trees. In pollen diagrams from sites throughout the country, there are falls in pine pollen values. In the Wicklow

6:4 A ring-barked pine. The technique is used to remove invasive pine species from native woodland. A similar technique may have been used to kill pine trees during the Bronze Age.

Mountains as in other uplands pine had already been in slow decline but the circumstances that led to the calamitous loss of the remaining pine some 4,000 years ago are not fully understood. A worsening climate had a part to play. Increased rain probably hastened the death of the pine trees as the species copes poorly with waterlogged conditions and, in soils already impoverished, pine seed germination rates would slump.

The deteriorating climate was not the only factor, for it was people with axes who also hastened the pine's demise. Unlike the felled hazels and oaks that re-sprout from dormant buds, if pine is cut down it dies, making it particularly vulnerable to the axe.

Fewer bog pine trees meant less water removed by transpiration through the pine's needle-shaped leaves, so soils became increasingly waterlogged and bog surfaces became wetter. The change in the character of the western uplands in particular was immense and in a few hundred years the mountain sides of the Nephin Beg range in County Mayo lost much of the dark, evergreen hues of their pines.

PAST VOLCANISM AND LINKS TO THE PINE DECLINE

There is a possibility that the loss of the native pine woods was hastened by very poor weather after a volcanic eruption far from Ireland. In the closing decades of the twentieth century, research has been carried out on the influence of volcanic activity on past climate. This research has linked information contained in the precisely dated Irish oak tree-ring record with ancient climate records from the Greenlandic ice sheet. Extremely narrow tree-rings in bog oaks point to times in the past when growing conditions for the trees were at their worst.

At times the dates of the narrowest tree-rings coincide with bands of volcanic acids trapped in the dated layers of the Greenlandic ice. This circumstantial evidence points to short periods of wet weather in the past forced by raised levels of volcanic acids and ash in the high atmosphere. It is known that the ash from Icelandic volcanoes can travel long distances in the wind. Iceland is over 1,000 km north of Ireland but microscopic traces of volcanic ash from a number of Icelandic volcanoes have been found in Irish peat bogs and lake sediments.

The volcano Hekla has erupted periodically since the last Ice Age ended. Precise radiocarbon dating of Irish peat containing its volcanic ash or tephra dated one of its eruptions, known as Hekla 4, to an age range spanning 2345 BC. At this date, just over 4,000 years ago, there is an extremely narrow tree ring event in the bog oaks. The combined evidence hints that the weather may have been even worse after the eruption.

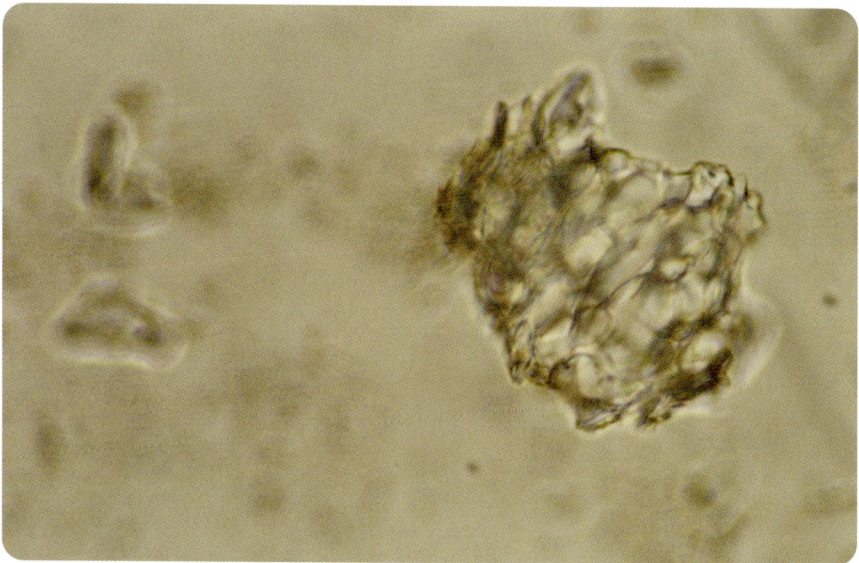

6:5 A microscopic volcanic ash shard from the eruption of Hekla, radiocarbon dated to 2395–2279 cal. BC

It was not only the land-grown pines that were dying out, for the bog pine forests were dwindling too. Pine can survive on a low level of plant nutrients but to do well the tree needs moderately high levels of phosphorus, potassium, calcium, magnesium, nitrogen and sulphur. If deprived of these essential nutrients, growth is retarded and the tree may die. Dead, land-grown pine timber decayed, releasing

6:6 A 4,000-year-old stunted pine trunk from Fallahogy Bog in County Londonderry

its nutrients back to the soil but the dead pines on the bogs were preserved by being engulfed in the peat. The annual growth rings in the bog pine timber are discernible so that the wood can be tree-ring dated. At Sluggan Bog dendrochronological studies show that bog pine had virtually died out by 4,500 years ago. The trees survived a little longer at some sites. At Fallahogy Bog in County Londonderry the pines that had lived for 150 years had trunks no thicker than an adult's wrist. Poor vegetative growth may have adversely affected the trees' fertility, as the low percentages of pine pollen in the pollen record appear to attest.

At Glashabawn in County Offaly the bog pines maintained a foothold for another thousand years, until at least 3,500 years ago. Stumps preserved in the peat from this site show that there were about 500 pine trees per acre and, like those at Fallahogy, the trunks were slender, being about 5–40 cm in diameter. It is possible that the last pines at Fallahogy, Glashabawn and also those on the Mizen Peninsula in County Cork were all growing close to starvation. The general picture drawn from the fossil record is of once-extensive pine woods dwindling to occasional patches of stunted trees.

Native Irish pine may not have died out completely for stumps dated to later times have been found, especially in the midland bogs. In its last strongholds, in Killarney and the Burren, pine lived on in places well away from waterlogged peat, where the tree survived as single specimens or in small copses. The last recorded native bog pine, a small tree that grew on Clonsast Bog, was radiocarbon dated to AD 134–650, although pine cones from Viking-age Dublin around AD 1025 and further evidence of pine living into the fourteenth century hint that the species survived in isolated places.

THE WIDER INFLUENCE OF THE PINE DECLINE

The loss of pine may have brought close to extinction the animal species that depended upon it. Some of the pine woods' insects were doomed whilst others that were more adaptable survived by moving into other types of woodland. Bird species were also adversely affected as their habitat fragmented or disappeared. The 'cock of the woods' or capercaillie (*Tetrao urogallus*), a large member of the grouse family, probably lived in the native Irish pine woodlands, as it does today in the remnants of the ancient Scottish pine forest, although Scotland's native birds were exterminated in the late eighteenth century. The modern stock originates from birds re-introduced from Sweden.

Where falling pollen percentages marked the pine decline in some Irish pollen diagrams, there is a rise in the levels of the pollen of ash (*Fraxinus*). Ash was the last native tree to spread and, as it is a colonising species, the rising values for ash pollen mark where woodland had regenerated, as it had close to Meenadoan Bog in

County Tyrone. Nothing in the pollen record points to there ever having been extensive ash woodlands in Ireland, but at Scragh Bog in County Westmeath the tree had ousted elm from its place as the major tree of the nearby woodlands.

THE SPREAD OF BLANKET PEAT

From the places where the woodlands were lost, there developed a new type of vegetation that was almost treeless. Where formerly there had been birches, pines and rowans, there grew mosses, heathers and sedges whose dead remains soon formed blanket peat. An understanding of where these plants may have originated explains in part how the landscape converted quickly to bogland. Plants that form blanket peat must be able to grow in wet conditions and withstand the temperature

6:7 A hummock of *Sphagnum* moss, surrounded by bog cotton leaves, growing on the surface of modern upland blanket peat

variations across the seasons. Such plants include *Sphagnum* moss species, heathers, sedges and purple moor grass (*Molinia caerulea*). Some of these plant species once lived in the moist conditions along the upland woodland edges and those that could cope with the open conditions after the woodlands were cleared formed the emergent blanket bog vegetation.

Delicate woodland moss species did not survive once their shady woodlands had gone because they dried out in the sun's heat, but *Sphagnum* coped with the rigours of open conditions because the moss can store water inside its leaves. *Sphagnum* leaves are composed of two types of cells: the narrow, living, photosynthetic cells and the larger, hollow, dead cells inside which water is stored, thus safeguarding the moss from death by dehydration. Water enters the cells through microscopic pores in the walls, which are strengthened against collapse by bands of supporting material.

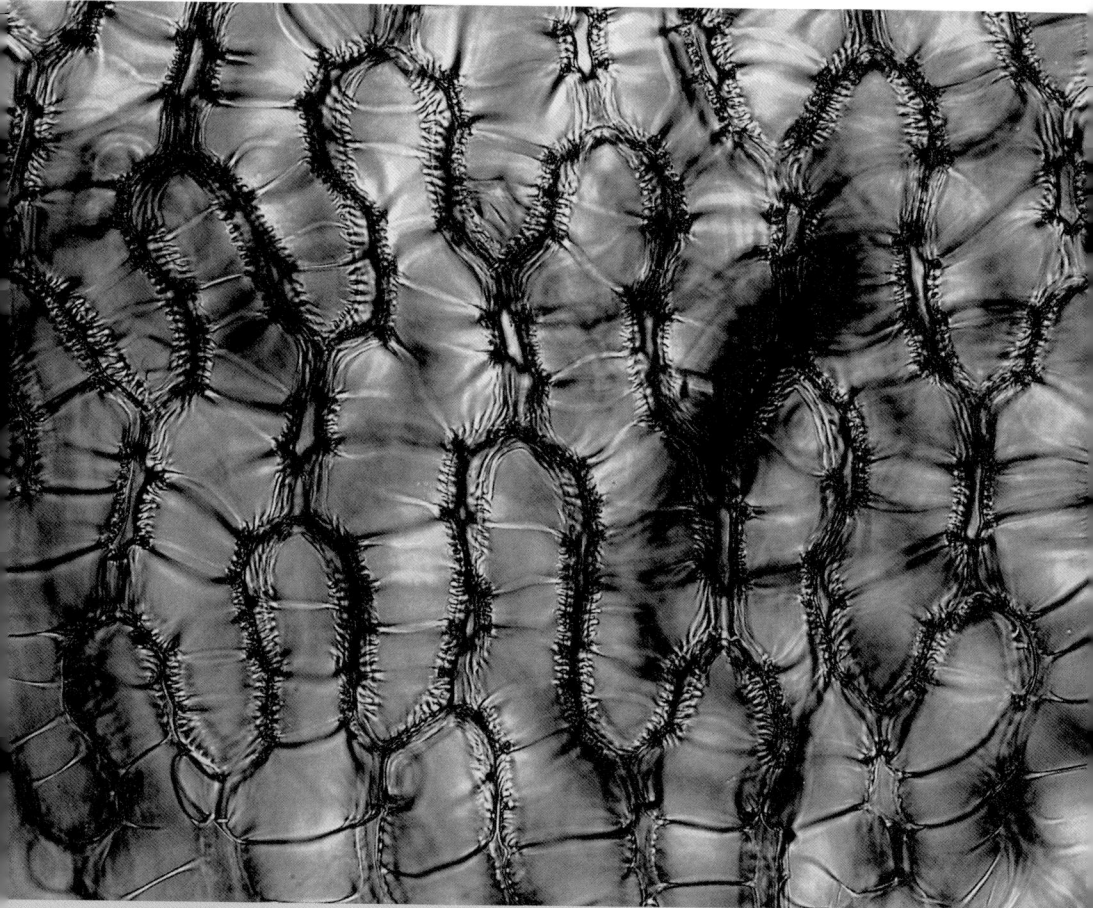

6:8 Photomicrograph of a leaf of *Sphagnum austinii* (formerly *S. imbricatum*) showing the large water-holding cells

Furthermore, all the nutrients that *Sphagnum* needs for growth are obtained solely from rain, making it the ideal plant for nutrient-depleted, wet places. Slopes that would otherwise have been well drained remained wet where *Sphagnum* grew. The moss soon spread out over the landscape and, as it died, underlying rocks or bits of dead trees were sealed beneath a thin skin of newly formed peat.

As the blanket bogs grew, they provided an enlarged habitat for plants that trap and digest small insects to meet their nutrient requirements. The protein these plants absorb from their prey is the source of essential nourishment. The insect-eating sundews (*Drosera* species) and butterworts (*Pinguicula* species) did well in the bog pools and on the sides of the moss hummocks that abounded in the blanket peat.

The colours of the former upland woods had changed with the seasons but this progression was less marked where upland peat prevailed. Apart from late summer, when flowering heathers turned the hillsides pink and purple, the landscape was dominated for the rest of the year by shades of brown. In a few places, near the tops of the high mountains as in the Mournes range, there remained a narrow, brighter, evergreen zone formed from crowberry growing in places where there was no lengthy snow cover.

There is no single set of conditions under which blanket peat forms – human activity and a worsening climate seem to act together in some places, although separately in others and the expansion of blanket peat was not uniform across the country. In the west in particular, blanket peat had begun to grow in previous millennia. There is blanket peat beneath the Neolithic Céide field walls, and in the south it is found below field walls on Valencia Island in County Kerry. Curiously, in the Mournes and Wicklow Mountains, blanket peat was eroding in some places whilst it was forming in others. The causes of erosion include disturbance by people and instability within the peat but there may also have been a climatic factor. At times improved weather let peat surfaces dry but when these dry surfaces were soaked during subsequent rainy periods, underground drainage channels formed, causing the overlying peat to collapse and disintegrate.

CONDITIONS AT LOUGH NEAGH

Lough Neagh is the largest lake in the British Isles and today it drains about 45 per cent of the north of Ireland therefore, in a sense, the lake acts as a giant rain gauge. It is considered by some specialists that the level of Lough Neagh's waters rose some 4,000 years ago in response to the increased amount of rain draining into the rivers that fed the lake. As the level of the lake rose, the expanding shallows along the lake edges provided the perfect environment for microscopic algae called diatoms. All freshwater lakes have diatoms living in them but rarely in such vast numbers that

6:9 The pale grey deposit beneath the dark brown peat is diatomite

their dead bodies form a deposit, as they did in Lough Neagh in the past. The algae are each encased in an outer wall rich in silica and, on death at the end of each growing season, the vast amounts of the outer cases sink to the floor of the lake and form a crumbly, whitish material. This rare deposit is called diatomite and it develops only in shallow, well-lit water where myriads of diatoms flourish.

Layers of diatomite within peat at Ballymacombs More, near the west bank of the Lower Bann River, appear to mark variations in previous lake levels, possibly in response to increased rainfall in the past. In contrast, however, there is evidence for drier periods from studies of the blanket peat in the Wicklow Mountains at about the same time as the diatomite was forming in the north. A combination of these contrasting sets of evidence makes it clear that the climate of the Irish mid-Holocene was not characterised simply by unremitting rain.

THE FURTHER SPREAD OF FARMING

An increasing population and its demand for food led to more land being needed for farming. The land farmed in the early Bronze Age did not overlap much with places cultivated in the Neolithic period, perhaps because earlier farming had robbed the land of its goodness. The heavy clay soils of parts of the north were still largely uncultivated in the Bronze Age and on these the densest woodland remained. From about 4,000 years ago most farming may have been on medium-quality soils that were less difficult to work and it may be that ease of cultivation mattered more than soil nutrient status. The best-drained lowland soils were preferred and people continued to live along the coast as well as beginning to farm the sandy, gravelly soils of the midland eskers. The pollen record for the opening landscape includes the appearance of the pollen of plantains and buttercups, classic indicators of pastoral farming.

A marked shift to stock breeding in the archaeological record at this time set the pattern for Irish farming ever since. A good example comes yet again from the Céide field system in County Mayo where, after a lull of about 1,000 years, cattle farming had resumed. In the County Mayo area almost 3,500 years ago there was a further advance in farming practice. The ancient practice of grazing stock on distant rough pasture during the growing season, thus protecting the cultivated land close to settlement, may have originated in this area. On the less well-drained land rushes and sedges may have grown and neither makes good grazing. Sheep and goats may have fared better than cattle on the poorer land.

Land was suited to grazing where grasses grew alongside plants from the clover family (Leguminosae). Clovers and their relatives, such as the bird's foot trefoil (*Lotus corniculatus*) have little bumps or nodules on their roots. These contain micro-organisms, often species of the bacteria *Rhizobium*, *Nitrosomonas* and *Nitrobacter*, which can trap nitrogen from the air and convert it into the nitrate that plants need to make their proteins. As an added benefit, when the plants die and decompose the soil nitrate level increases.

FOOD AND THE NEW FARMING

The point at which people in Ireland became entirely dependent on farmed food is imprecisely dated because there is no clear-cut evidence for complete abandonment of former food-collecting practices. Agriculture first began over 5,500 years ago and the process of change from food-gathering ways to farming took almost 2,000 years, by which time the entire country was inhabited. There is evidence that some of these inhabitants continued to gather at least a proportion of their diet. The people living at Whitepark Bay in County Antrim left a rubbish heap, known as a kitchen midden, of bones and shells. Amongst the bones of farmed animals such as cattle, pig, sheep or goat, and horse, there were those of red deer, wolf or dog, geese and gulls mixed with cod bones and shells of limpet, periwinkle, dog whelk, mussel, oyster, cockle and razor shell. Here at least, farmed and wild-gathered foods provided a varied diet. Nevertheless, by 3,500 years ago, most experts agree that Irish people relied primarily on farmed food.

6:10 A charred cereal seed from a Bronze Age archaeological context

A reliable source of food such as cereals might be considered nourishing and healthy but Bronze Age human bone evidence shows otherwise. Skeletons of children living at that time show signs of anaemia, a blood disorder that could have resulted from a diet overly reliant on cereals, from which the human body has difficulty in absorbing enough iron for its needs. Evidence for the disorder was found on a child's skeleton from County Fermanagh. People may also have been weakened by internal parasites like flukes and tapeworms, picked up from close proximity to cattle, sheep or pigs and, where they lived under poor sanitary conditions, infectious diseases spread. Poor nutrition doubtless worsened the effects of gastro-enteritis and other bacterial infections.

The mainstays of stock farming at this time were cattle, pigs, sheep and goats, these latter animals having been introduced during the Bronze Age. Bones of farm animals showing butchery marks turn up at archaeological sites from this period, although some cattle may have been kept for milking. Doubtless sheep were farmed for their meat and then their wool and the horse may have been put to various uses. The earliest horse bones were found at Newgrange in County Meath and date to about 2400 BC. Early horse remains are extremely rare in Ireland. Old horse bones were found with food refuse so it seems that they were eaten after a life of being ridden or pulling a plough although there is no wear evidence on the teeth to suggest the use of a bit. As stock farming expanded, the flocks and herds were at the mercy of bears and wolves. Bears lived in Ireland until about 3,000 years ago with some of the last surviving in County Leitrim. Wolves lived on much longer and were finally wiped out near the end of the eighteenth century. It is claimed that the last wolf was killed at Myshall, close to Ballydarton, County Carlow about AD 1786.

Apart from the occasional broken stone ard point, the tools used to prepare the land for crops rarely survive in the archaeological record. At an ancient field system at Carrownaglogh in County Sligo, grooves made by an ard were seen on the surface of a prehistoric field, where the soil had been broken before cereals were sown. Other fields had their soils ridged, although not necessarily to improve drainage because the ridges ran across rather than along the slope. Barley and, increasingly, wheat, with a little oat (*Avena* species) and rye (*Secale*), may have been grown in rotation in the small fields. Some land was left fallow to improve grazing but not where bracken (*Pteridium aquilinum*) invaded the resting land.

Finds of the bronze sickles comprising a curved knife with a wooden handle that can be held in one hand and used in harvesting are a little more common. To reap the crop one hand held the sickle and the other grasped the cereal stalks below the ripe seed-heads. The sickle blade was used to slice through the stalks. In a field of ripe grain, it would be hard to avoid including the tops of the taller weeds and so the harvested cereal crop may have contained seeds from charlock (*Sinapis arvensis*), knotgrass (*Polygonum persicaria*) and corn spurrey (*Spergula arvensis*). Grain thus became contaminated with weed seeds, which may not have been the case in earlier times.

A NEW AGRICULTURAL CROP

The date of the first flax (*Linum usitatissimum*) crop grown in Ireland is contentious and the pollen record for its introduction is ambiguous because there are small native flax plants of no economic worth that produce pollen grains indistinguishable from those of the cultivated species. Where the first pollen records for flax are set within an agricultural pollen profile, then one may have a measure of confidence that flax was grown, such as on the land near Essexford Lough in County Louth.

The flax seed record from archaeological sites is also fairly reliable. There is an impression of a flax seed on an Early Bronze Age pot from Agfarell in County Dublin. Both records may point to the good land near the east coast used to cultivate a crop grown for its fibre or its nutritious seeds. Flax seeds in the archaeological record are usually carbonised but the charring process shrinks them, making it difficult to distinguish between ripe, shrunken seeds and the slightly smaller, immature seeds from flax grown for fibre. From the archaeobotanical evidence alone, it is not clear which use was commoner.

When grown for its fibres the plant was pulled, not reaped like a cereal. The flax plant's fibres stretch from the tip of the stalk to the roots, so pulling up the whole plant makes the most of the valuable crop. Textiles were made during Bronze Age times in Ireland and there have been occasional finds of textiles made from spun

6:11
The characteristic blue flowers of flax

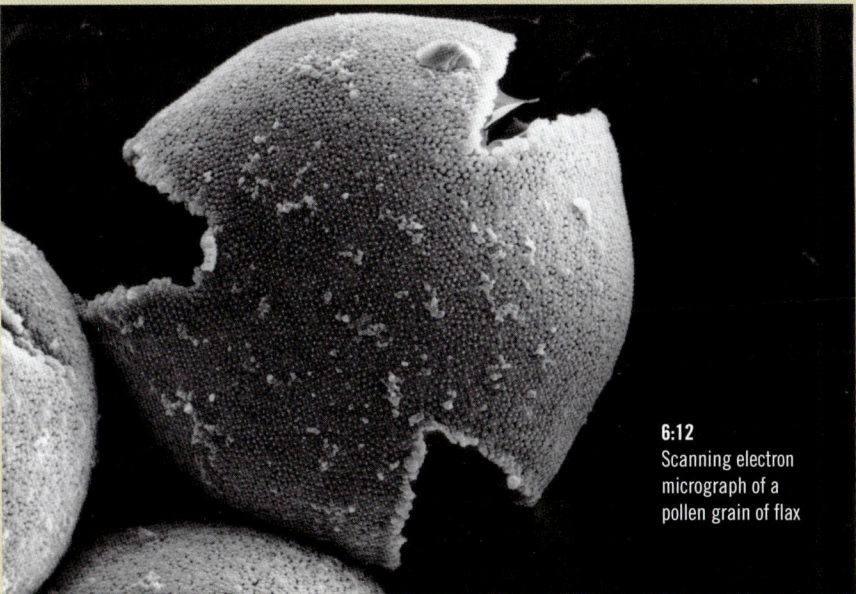

6:12
Scanning electron micrograph of a pollen grain of flax

and woven animal fibres, for example, the horse hair belt from Armoy in County Antrim. Protein-based fibres preserve better than the cellulose-based plant fibres in wet, oxygen-depleted soils and although Irish soils do not preserve flax fibres well, these may have been spun and woven into linen.

TRACKWAYS

At a number of sites prehistoric bog trackways of various dates have been found. One trackway, dated to 2500 BC, revealed that in skilled hands a stone axe was capable of sophisticated woodworking techniques. Stone axes were still in use at a time when metal axes had come into the country. Somewhat later, about 2259 BC, oak timber from a trackway carries metal tool marks. A variety of construction methods were employed, dictated in part by the wetness and softness of the bog beneath. One trackway was made of a bed of birch and ash runners with round-wood stems of alder about 2 m long lying on top. Pegs of alder and birch were used to pin the whole structure to the bog. Others of same age were made mainly of oak and ash. All are fine examples of the ingenious use of local resources and skills.

Throughout the period from 4,000–2,700 years ago, the pressures placed by people and climate change on the landscape increased steadily although the timing and intensity of impact varied from place to place. The period 4,000–2,700 years ago is characterised by an opening of the landscape, the expansion of farming and the general demise of woodland, yet the pollen record from Red Bog, in County Louth, and Mooghaun Lake, County Clare, showed widespread woodland regeneration and an overall demise in the pollen record for crops and weeds. The Late Bronze Age is separated from the Iron Age to come by a diminution of archaeological material and the landscape's evidence for lessened human pressure. During the remaining part of the first millennium BC and the first millennium AD the landscape underwent further changes that included settlement expansion as well as abandonment of some parts of the landscape and the rise of a new type of intensive cattle farming.

CHAPTER 7

THE LANDSCAPE IN LATE PREHISTORIC AND EARLY HISTORIC TIMES

2,700–1,000 years ago

TIME PERIOD FOR
THIS CHAPTER

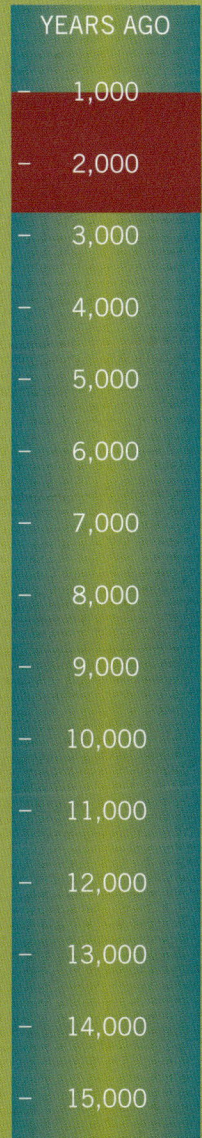

KEY ISSUES:

the last of the bog oak woodlands;

bog trackways;

bog bodies;

dairying;

the last expansion in woodland;

historic records;

ringforts;

monasteries.

YEARS AGO
1,000
2,000
3,000
4,000
5,000
6,000
7,000
8,000
9,000
10,000
11,000
12,000
13,000
14,000
15,000

In the earlier stages of this account of the landscape's history, major developments brought about largely by natural forces have been described, followed by a strengthening of the impact of human activity on the environment. Further evidence for landscape change, beginning in the first millennium BC, comes from the fossilised timbers of the last of the bog oak woodlands, wooden trackways and, spectacularly, the bodies of people, all embedded in the peat of the lowland bogs.

THE LAST OF THE BOG OAK WOODLANDS

Bog oak woodland had almost disappeared from Ireland by 2,200 years ago (200 BC), thereby making those that remained north of Lough Neagh and at Garry Bog in County Antrim the last of their kind.

What had happened to bring these ancient woodlands close to extinction? There is no single answer to this question. Starvation of the oaks may have played a part in the woodlands' demise. In earlier times bog oak trees had supplemented their food supplies from the rich inorganic sediments beneath the bog, into which their roots had penetrated. As peat deepened with the passage of time, the trees' roots could no longer reach that source of food and so the trees starved. Poor feeding doubtlessly affected the trees' ability to produce acorns and across the country the bog woodlands dwindled as old trees were not replaced by new saplings.

The already failing health of the bog oak population may have further weakened during very wet weather. The bog oak timbers from Garry Bog have an extremely narrow annual ring dated to 207 BC, thought to be the result of very poor growth when the bog became exceptionally wet. A distant volcanic eruption could have forced increased rainfall if vast quantities of acidic gases and dust had entered the

7:1 A slice through the trunk of one of the last Irish bog oaks

high atmosphere. After the eruption of Mount Pinatubo in the Philippines in 1991, the autumn in Ireland in 1992 was very wet and perhaps something similar happened about 207 BC, forcing death upon the last of the bog oaks.

BOG TRACKWAYS

Additional indirect evidence of a general worsening of the weather during the closing centuries of the last millennium BC comes from big wooden trackways embedded in the peat of Corlea Bog in County Longford. It is thought that the large trackways were built to let people cross wet bogs that had been, until that

7:2 Iron Age trackway of substantial oak planks at Corlea Bog in County Longford, dating to 148 BC

time, reasonably safe to traverse. Older trackways, of which at least 200 have been identified, received comment in previous chapters of this volume but none of these were as large or as well built as the Iron Age trackways at Corlea.

It took about three hundred well-grown oak trees and a similar number of birches to make just one of these 'wooden roads'. The removal of so much old-growth oak and other tree species from the local woodlands must have adversely affected their ecology. The size and number of the oaks felled has been gauged from the length and dimensions of the timber from one trackway, in which most of the planks were 3–3.5 m long and 15 cm thick. Planks that size are heavy and, where the peat was particularly soft and yielding, brushwood, reused timbers, woodchips, and runners fixed with wooden pins all helped to stop the planks sinking into the bog. One of the trackways was almost 1 km long and it gave dry footing to people and possibly domesticated animals but not, it seems, to carts or chariots: the trackways' planks show no wear marks from wheeled vehicles. Building the Corlea trackways needed a big workforce, thus providing indirect evidence for substantial numbers of people in this part of the country. Bog trackways continued to be made until the sixth century AD but, when compared to those at Corlea, they are simple structures with some composed of little more than a line of single planks.

BOG BODIES

Of all the materials from Irish peatland that have informed this story few are as spectacular as the bodies of people. The manner in which peat preserves human remains is not fully understood but it is clear that microbial decay is almost halted by the cold, acidic, oxygen-depleted bog water. Few of the bog bodies found have been scientifically investigated and none have been studied as thoroughly as the two found in the Irish midlands in 2003. In February of that year, at a bog at Clonycavan in County Meath, the head, partial torso and parts of the arms of a man were found.

Then in May the upper torso, arms and hands of another man were found in a bog at Oldcroghan in nearby County Offaly, 40 km from Clonycavan. These finds bring to approximately five the number of Irish bog bodies dated to the Iron Age. Oldcroghan Man's body was radiocarbon dated to 361–115 cal. BC. Entombed in peat for over 2,000 years, the flesh of the bodies remained as soft as in life and their skins were stained deep brown from centuries of immersion in bog water. On investigation the bodies of both men showed signs of violence. Clonycavan Man's nose had been broken, his skull shattered and his abdomen sliced open, whilst Oldcroghan Man had been beheaded and withes forced through his upper arm muscles. That both men had met such violent deaths led experts to believe that they had been ritually slaughtered, although murder cannot be ruled out.

Above: **7.3** The head and upper torso of a man's body found at Clonycavan in County Meath

Below: **7.4** The upper torso, arms and hands of the man found in Oldcroghan in County Offaly

Most of the bog bodies discovered in recent decades have been pulled out of the peat or have been found in peat excavation machinery so it is rarely that environmental reconstructions can be attempted from the deposits that had surrounded them. Because some of Oldcroghan Man's fingernails were found embedded in the peat, it was possible to pinpoint where his body had lain. A vertical column of peat that spanned the location of his fingernails was the basis for the environmental reconstruction based on studies of fossil pollen, *Sphagnum* leaves, beetle fragments and microscopic single-celled animals known as testate amoebae.

The tiny amoebae live in the upper few centimetres of the bog vegetation, where some species can tolerate some dryness whilst others thrive only in the wettest conditions. Much can be learned about bog surface moisture from the study of the little amoebae's distinctive outer cases. The testate amoebae record proved that a pool had been forming in the bog at the time when Oldcroghan Man was alive; this finding was supported by leaves of *Sphagnum* species that also grow only in bog pools. Proof that the man's body had been plunged into the bog at the time of his death and not left to rot on the bog surface came from fossilised beetle fragments, amongst which no species that lived on decomposing flesh was found.

The pollen analytical investigation showed woodland clearance and an increase in human activity on the dry land around the bog at the time when Oldcroghan Man died; a shift in land usage that has also been identified to this time at other sites. There was elm pollen in the peat profile, showing that this tree grew in the woods of County Offaly during the first millennium BC. This interesting discovery contributes to the history of this native tree, often considered as uncommon in Irish woodlands since Neolithic times.

SCRUBLAND AND CATTLE

The scrubby grassland that remained after trees were felled was the mainstay of cattle farming and it was managed so that there was fodder for the animals throughout the year. Knowledge of local soil and drainage conditions was essential for maintaining food stocks, therefore management practices could not have been uniform across the country. In many areas, during the spring and summer, the cattle were probably taken away to pasture. Little is known about how the practice of taking animals to more distant pasture in better weather began but one reason why this practice may have been used was that it made the most of the less nutritious grasslands. The cattle could eat the newly-sprung, tender shoots of otherwise tough and unpalatable grasses, keeping the adults fed and their calves supplied with milk. In autumn, on return to home grazing, there was fresh grass to be had as this sprang up in the stubble once the cereal crop had been harvested.

7:5 Fresh grass foliage springs up between the stubble stems in the warm moist days of autumn

On the better soils hulled and naked barley as well as wheat were probably grown in big patches, with little to hint that any arable land was enclosed. In preparation for sowing, the soil was turned and seed scattered once the soil surface had been dried by the spring winds. The crop grew as the spring and summer days lengthened, the soil being moistened by frequent rain. The ripe crops were harvested in the late summer or even early autumn, as the timing of the harvest depended on a period of dry weather. After harvesting, moisture and light reached the grass and weeds previously shaded by the cereals, and soon the crop's weedy understorey sprang up, providing fresh fodder for the cattle. Many of the weeds of the cereal fields were the same species as those found in pasture. When the fresh growth was grazed, animal dung dropped and trampled into the soil maintained fertility for the following year's crops.

By the onset of winter the number of animals in the herd had been reduced because beasts were culled in the autumn. As each year drew to a close and the weather worsened, the herds-people's skills were vital for keeping the remaining animals fed. Hay was not harvested in prehistoric times so winter fodder remained as standing grass stems, beneath which leaves continued to grow in all but the harshest weather. Managing cattle on land that remained firm during the winter was less difficult than where animals were kept on poorly-drained, soft soils, where pasture was vulnerable to trampling.

Where cattle, sheep or goats were fed and then taken off the land, its nutrients became depleted, thereby creating ideal conditions for the flowering plants that did not thrive on rich soil. On the grasslands of the better-drained drumlin clay, there bloomed the yellow daisies of the hawkbits (*Leontodon* species) and purple-flowered knapweed (*Centaurea nigra*). On damper soils there were plenty of green-stemmed rushes and sedges (*Juncus* and *Carex* species) growing amongst the grasses, with the countryside of the wet west smothered in spring with marsh marigold's (*Caltha palustris*) yellow flowers. At Céide Fields, the cradle of Irish cattle farming, once again cows fed on grassland in which buttercups (*Ranunculus* species) and clover grew abundantly. Fragments of flower-filled meadows like those described are still to be found today and some may have remained largely unchanged since prehistoric times.

7:6 A meadow whose vegetation may resemble those of late prehistoric times

PRODUCTS OF IRON AGE FARMING

There is little archaeological evidence for settlement during much of the Iron Age and this may point to people having become largely nomadic, with the care of cattle central to the lives of most. Cows eat grass but it is people who decide how animal products will be used. Animal bones from archaeological sites prove that cattle were more common than pigs. Pigs are a 'single use' animal, providing only meat, but cattle gave both meat and milk, the return of energy and nutrients from milk outweighing that from beef. The history of Irish beef farming may be as old as the first cow kept but the date when dairying was first practised has experts wrangling, with some arguing that it was rare in Iron Age times.

Surprisingly it is the bogs that hold some evidence for dairying, in the form of 'bog butter'. Chemical tests have shown that the lard-like substance remaining after fat had been long immersed in bog water was once butter. Caches of 'bog butter', some weighing over 20 pounds, have been found buried in wooden containers in bogland throughout the country. Samples of bog butter have been radiocarbon dated and the age span for the greasy material runs from 400 BC to the thirteenth century AD. Most dated bog butter had been made between 375 BC and AD 313, lending strength to the argument that dairying was practised during the Irish Iron Age, but it could also be argued that the Iron Age bog butter represents nothing more than the seasonal exploitation of milk. The point in time when cattle were farmed for milk rather than beef would mark the start of dairying but to ascertain if it had been practised in the later Iron Age, one would need to know the gender balance in the herds. From 100 BC to AD 600, however, the domesticated animal bone record is non-existent and it is also worthy of mention that the tree-ring record is skimpy from 40 BC for the next 350 years, all because of the lack of archaeological sites. The point in time when dairy farming in Ireland began eludes us still.

The evidence for cereal farming is not strong for any part of the country throughout the Irish Iron Age but there is enough information from the pollen records to show that it never died out, although a nomadic lifestyle is not best suited to cultivating the land. Records of cereal pollen in pollen profiles serve as a reminder that even if most people's everyday experience was with the herds, some others cultivated and harvested crops.

THE LATE IRON AGE LULL: THE RETURN OF THE WOODLANDS

In the last centuries BC and the first centuries AD, from the limited number of sites that have yielded palynological evidence, is seen an upsurge in the tree pollen percentages at the same time as the farming signal dwindled overall and this is matched by an overall thinning of the archaeological record. Professor Frank

Mitchell used the term 'Late Iron Age Lull' to describe this period of much reduced human activity. The dating at some sites is better than at others and, where time frames are secure, the pattern in the pollen record across the regions can be traced. It is puzzling, however, that in the last century AD there is dendrochronological and archaeological evidence for the building of high status structures simultaneous with pollen evidence that indicates partial abandonment of land.

The record from Beaghmore in County Tyrone, between 100 BC and AD 350, is characteristic of sites throughout the country where earlier clearance was followed by woodland renewal. This secondary woodland, as it is known, tended to be dominated by trees that can grow rapidly. In the north, the heavy clay soils favoured hazel, holly and birch. On the Dingle Peninsula the new woods had plentiful birch, alder, willow, hazel and holly, whereas in the west the newly expanded woodland was dominated by yew.

This trend toward re-invasion by trees was, however, not universal. At Emlagh Bog, the landscape remained open and low-level land use continued, as it did at Clonfert Bog in County Galway. In even stronger contrast, at Red Bog in County Louth during the first century AD woodland was cleared and farming was thriving, as was also the case in northeast Mayo, where arable agriculture was re-established by AD 300–500. Where the western soils were disturbed by renewed farming, old soil humus was washed into nearby lakes and incorporated into their basal sediments but the pattern is not universal as nothing similar happened at Lough Neagh at any time between 2000 BC and AD 1700, possibly indicating that farming affected soils less aggressively in this part of the north.

Theories have been developed to explain the apparent lack of people and a subsequent return of woodland, with some experts arguing for a massive population collapse. Others have asserted, with appeal to the same body of evidence, that there was no loss of population and that the archaeological and environmental records would inevitably be almost silent if people were living nomadic lives. As the bone and dendrochronological records have been hard hit by the lack of settlement sites, it is the pollen record that shows where there were people on the landscape. The pollen record is partially supported by one of the earliest documentary accounts of the Irish landscape. In a commentary by a Roman writer there is praise for the good grazing but no mention of a land devoid of people.

FROM PREHISTORY INTO HISTORY

This is the point in time where the records embedded in the landscape are joined by the written record, thus marking the divide between prehistory and history. The early writings contain much that is well dated and rich in detail but contain little about the landscape. *The Annals of the Four Masters*, however, make mention of

7:7 The memorial to the Four Masters, Kinlough, County Leitrim

some exceptional environmental events. The *Annals* are a commitment to the written word of an oral tradition spanning more than a thousand years. Written in the seventeenth century, they comprise a year by year account of important happenings, beginning early in the first millennium AD. Within the record are comments about years when the weather was particularly harsh or when the acorn crop was especially heavy and, overall, this additional information enriches the landscape's history.

THE SIXTH CENTURY AD, A TIME OF FURTHER CHANGE: THE AD 536 EVENT

Earlier parts of this account explained what happened to trees on bogs when they lived under conditions of environmental stress. It is especially notable therefore that the most widespread narrow tree-ring event in the 7,000-year-old Irish tree-ring record did not show in bog oaks for, by this time, bog oak woodland had become an ecosystem of the past. The tree-ring event occurred in oak trees that grew on fens as well as those presumed land-grown and therefore well nourished and somewhat buffered from environmental stress. The tree-ring record shows an event marked by a narrow ring starting in AD 536, with the narrowest ring dated to AD 540–41. This narrow ring event is recorded in timbers from Ireland, Europe, eastern Asia and the western United States, meaning that some force was strong enough to limit tree growth across much of the northern hemisphere.

The event lasted in excess of six years, which is exceptionally long when compared with the three years of poor growth attributed to climatic downturn presumed to have been forced by volcanism. Even the huge eruption of Tambora in AD 1815, one of the largest in the last 11,000 years, did not damage the climate for six years. There is further evidence of disaster in the written record for Ireland for that time. *The Annals of Ulster* documented 'failure of bread' in AD 536 and 539, thereby linking a very early written account to one of the most puzzling in the tree-ring record. Could it be that the same cause forced the famine and the poor growth in oak trees? Is it possible that two huge volcanic events happened almost simultaneously, causing climatic deterioration lasting longer than any recorded previously in the tree-ring record? Unfortunately, the Greenland ice cores that keep a record of ancient volcanism and atmospheric dust do not provide an unambiguous answer.

Or could there have been a different cause? It has been suggested that the extended tree-ring event may be a response to poor weather after an extra-terrestrial incident. The Earth may have passed through a cloud of interstellar dust or a comet may have plunged into one of the world's oceans. Could the vast amounts of dust and water vapour shot in the high atmosphere after cometary impact have mimicked volcanic conditions and have caused a climatic catastrophe? As yet, no straightforward answer is forthcoming.

RINGFORTS: THE NEW FARMSTEADS

There were other swift and far reaching changes in the Irish landscape that also began early in the sixth century AD. In the archaeological record suddenly there is evidence for people almost everywhere except in the high uplands as new farmsteads were built. Today the remains of these farmsteads, as mounds surrounded by one or more circular banks and ditches 30 m in diameter, may be seen all over the country.

The ringforts or *raths*, as the farmsteads are variously known, were not restricted to any one area, soil type or altitude but fewer were built in the northwest than in the north midlands. It has been estimated that between 30,000 and 50,000 ringforts were built in the period AD 550–850 and the total sum of the Irish ringforts in Ireland is greater than all of the settlement sites recorded for Western Europe at this time. They have no exact parallel anywhere else, possibly because Ireland alone had the combination of soil and climate that favoured lush grasslands and cattle farming. Considering that the Iron Age population is almost invisible in

7-8 Aerial view of the banks and ditches of the ringfort at Cloncannon in the Barony of Ikerrin, County Tipperary.

7:9 *Raths* or ringforts may be identified by an outer bank and ditch or by a raised platform as at Ballymacarron Rath in County Down

the archaeological record, it is amazing that in the early Christian or medieval period thousands of the new farmsteads were built in only about 300 years. Their rapid construction in such large numbers must have taken a considerable workforce.

Until recently it had been thought that ringforts were connected to the introduction of dairying, possibly from Romanised Great Britain and Europe, as the rath was thought intrinsic to dairy cattle management. Now that it has been surmised that dairying could have been practised during the previous 700 years, a reassessment of the function of the ringfort is needed. The huge numbers of ringforts must represent some measure in the expansion in dairying although archaeological excavation has shown that some raths had other uses. Some experts assert that ringforts were defensive structures and many were, but primarily for the guarding of cattle rather than people. If the elaborate cattle raids described in ancient tales such as *The Táin* or *Cattle Raid of Cooley* are true reflections of the extent of thievery, then gathering the farm animals into the ringfort enclosure at night would safeguard them from raiders. In spite of their large numbers excavated ringforts are few, but the one at Deerparks Farm in County Antrim yielded bodies of parasitic lice of cattle, sheep and pigs as well as dung beetles, proving that the animals had been confined within the ringfort, although exactly where and how is not known.

There was also a re-establishment in cereal cultivation but on nothing like the scale of that experienced by husbandry. At some point, probably in the early medieval period, a new plough was brought into Ireland. A well as having an efficient blade called a coulter that cut into the soil, it had a device known as a mouldboard that turned the soil over so that the surface weeds were buried. Few people may have had access to the new technology, since soil preparation with the ard continued, as it had for centuries. Oats were once thought to have been introduced to Ireland in the early centuries of the first millennium but recent studies prove that this crop had been grown in Ireland since Bronze Age times.

CHRISTIAN BUILDINGS AND PRACTICES

The first ringforts coincided with the arrival of structures and practices unique to Christianity. The number of early monastic sites is miniscule when compared to the plethora of ringforts but the influence of farming practices introduced along with Christianity must not be underestimated. The honey bee (*Apis mellifera*) is probably native but bee-keeping appears to have expanded during the late fifth or early sixth century AD, possibly linked to the new monasticism.

The needs of the monasteries for writing materials were served well by the large cattle herds as the young calves' hides were used to make vellum. There are contradictions here, as a farmer's produce could suffer if calves were slaughtered young because without her calf beside her, a cow would not let down her milk. Perhaps the best that can be said of these seeming contradictions is that a compromise was reached that satisfied both the needs of the farmers and the requirements of the monks.

From the seventh and eighth centuries, our understanding of Irish farming practices is revolutionised by the extraordinary wealth of information in the law texts known as the Brehon Laws. The texts describe the care of domestic animals and their products such as hides, dung and tallow. Other than through the written record, there would be no way of knowing that sheep were milked or used as currency. Goats were not considered important but pigs receive frequent mention. Horses and their care and usage are described fully but because

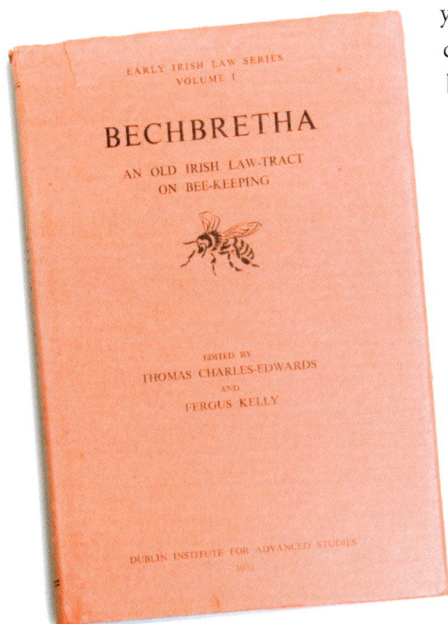

7:10 The bee laws governed bee-keeping practices

7:11 The donkey may not have been used as a beast of burden until the opening years of the second millennium

the donkey receives almost no mention it is thought that it was not a common beast of burden in the later part of the first millennium AD. As well as domesticated mammals, the law text state that hens, ducks and doves were kept, yet the archaeological bone record contains limited evidence of domestic fowl at that time.

There is information on growing peas and beans as well as crops that seem to be celery and leeks. The law texts recommended that by using a system of mixed cereal cropping and rotation the best could be obtained from the cultivated land. There are strong indications that this much revised arable system was brought from continental Europe or Great Britain by returning monks, as the words used had been derived from Latin.

From the seventh century land fertility was maintained or improved by dunging as well as by dressing with lime-rich marl or 'corn gravel', as the material is known in some parts of the country. The limited pollen record for cereals may hint that there was no great increase in the amount of land put to the plough. At this point in our narrative it behoves us to remember that the pollen evidence for arable agriculture comes from bogland far from places where oats, wheat and barley were grown. The intermittent cereal pollen evidence from the ninth century AD for some parts of the country does not demonstrate that cereals may have been grown extensively, but at that time grinding on quern stones had given way to the horizontal mill. This type of mill could cope readily with copious grain and most

7:12 Timber from a horizontal mill

of the timbers in the dendrochronological record for this period come from excavated horizontal mills.

It has been suggested that the preponderance of horizontal mill timbers in the tree-ring record point to a landscape in which oak may have been at a premium and its timber possibly reserved for agricultural use. There are signs of burgeoning agriculture in the ninth century along the valley of the Lower Bann River in the north of Ireland. A tephra layer, dated to approximately AD 860, has been found in peat from bogs on both banks of the Lower Bann and it marks a point in their pollen records that showed a landscape much denuded of trees and where cereals and flax were flourishing. This trend is not uniform across the country but the example from the Lower Bann valley reminds us that, at a time when the population was rising and the need for food was increasing, timber may not have been plentiful.

As the first millennium AD drew to a close and some of the great monasteries began to go into decline, the arrival of the Norse people, known now as the Vikings, brought terror as well as trade. Excavation of their towns, often located to give ready access to the sea, show that they enjoyed a lifestyle in which exotic produce was available alongside traditional fare. Fragments of the shells of walnut (*Juglans regia*), a tree not native to the British Isles, were found in the excavations at Fishamble Street, the site that marks the location of Viking Dublin. The excavations also revealed silk head-coverings, amber and walrus ivory, all of which point to successful trading with merchants in distant lands.

Although substantial Irish oak timber may have been at a premium in the ninth century, by the eleventh century some good timber must have been available. A huge Viking longboat, found in the Roskilde Fjord in Denmark where it had lain for almost one thousand years, had been built from Irish oak. Astonishingly, the boat timbers matched the Irish rather than the continental tree-ring record proving that the craft was built from oak trees hewn in AD 1042.

From the eleventh century onwards the Irish landscape underwent further changes, some brought about by military needs and others brought about by the introduction of new crops and all played out against a changing climate. It is recorded that Europe's weather had improved at the end of the first millennium AD but it is still remains unknown when the climatic improvement began. Some authorities place the change at AD 850 while others would argue for a date around AD 1050, but regardless of when the European weather began to warm, it is not clear if Ireland felt much benefit. The story in the chapters that follow is one of wars and agricultural imports that altered the Irish landscape still further.

CHAPTER 8

THE ORIGINS OF THE MODERN IRISH LANDSCAPE

1000–100 years ago

TIME PERIOD FOR
THIS CHAPTER

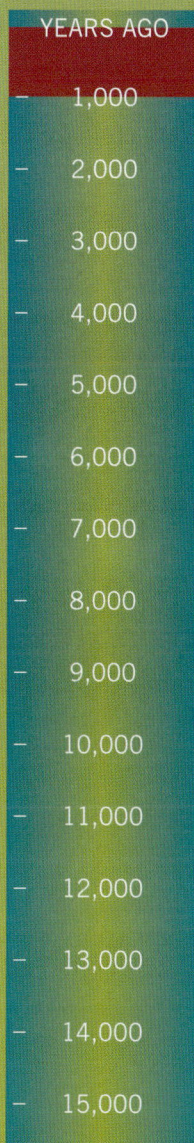

KEY ISSUES:

changes in bog flora;

historically dated Icelandic volcanism;

changes in farming practice;

the influence of the Black Death in the
fourteenth century;

woodland and scrubland loss;

increase in the cultivation of potatoes;

the enclosed landscape;

potato blight and the Great Irish Famine.

YEARS AGO

1,000
2,000
3,000
4,000
5,000
6,000
7,000
8,000
9,000
10,000
11,000
12,000
13,000
14,000
15,000

Many earlier chapters of this narrative opened with a description of the features in peat that can be seen with the naked eye, but the change in the peat that accumulated as the second millennium AD began is best discerned using a magnifying glass or a microscope. Until approximately the early tenth century AD the main bog-building moss was *Sphagnum austinii* (formerly *S. imbricatum*), a robust species that forms large hummocks. Until then the dead remains of this moss had built most of the vast lowland peat mass. At some time between AD 1000 and 1400, *S. austinii* was replaced by other *Sphagnum* species, in particular *S. papillosum* and also *S. magellanicum*, a species that today occurs on every continent except Antarctica. Climate change, falling bogland water tables, competition from other moss species and changes in the usage of land surrounding bogs may have contributed to the great bog-builder's demise. Today the moss is mostly limited to the bogs in the west of the country.

In the peats that formed from the twelfth century onwards, there are more layers of microscopically thin Icelandic volcanic ash. These layers have improved the dating precision for studies of the most recent peat, since tephra layers from the eruptions of the Icelandic volcano Hekla in AD 1104, AD 1510 and AD 1947 as well as the AD 1362 eruption of Öræfajökull have been confirmed.

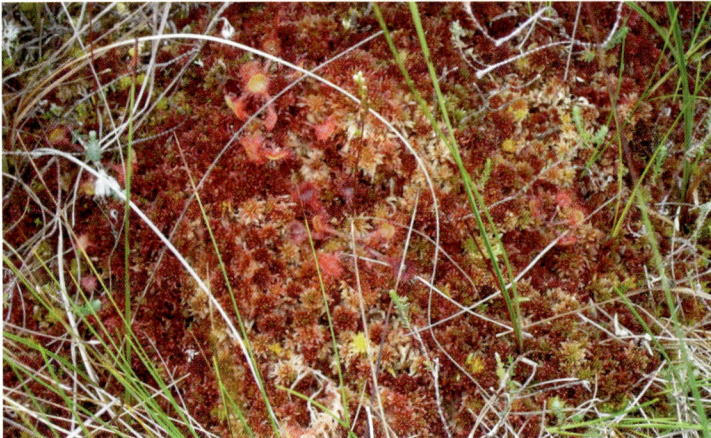

8:1
Sphagnum austinii hummock on a western bog

RESURGENCE AND CHANGE IN FARMING

Historians assert that from the mid-twelfth century the Anglo-Normans influenced Irish agriculture by introducing practices such as hay-making, but as hay is difficult to make in a country where rain is an almost daily occurrence, it is likely that hay-making was not widespread. Although the Anglo-Normans have been credited with the introduction of soil improvement practices, these are of greater antiquity in Ireland and it may be more accurate to say that the practices expanded from the

8:2 New monastic orders such as the Cistercians established foundations at many locations, including Abbeylara in County Longford (left) and at Greyabbey in County Down (above). The monks may have introduced and encouraged innovative agricultural practices

twelfth century. Marl, as well as shell-rich shore sand, was used to improve soil fertility and structure, with the sands brought inland from coastal areas that included Cork harbour, Kinsale, Youghal, Tralee and Lough Foyle.

It has also been said that it was the Anglo-Normans who encouraged a greater emphasis on cereal farming but some pollen profiles that incorporate the AD 1104 Hekla tephra layer show that the cereal record had strengthened before the Anglo-Norman advance of AD 1169, meaning that the increase in cereal farming may have been in response to local need. The early eleventh century archaeological domestic bone evidence may hint at why more cereals were grown. At Knowth in County Meath the bones found show a reduction in the percentage of cattle bones and an increase in the percentage of those from pig, reflecting the reduced importance of cattle-keeping to the agricultural economy that is discernible throughout the faunal record at the end of the first millennium. Is it possible that the change-over to pig farming and the increase in cereal cultivation may be linked? Pigs may be fattened on acorns and the *Annals of Ulster* record great harvests of oak 'mast' (the term used for an acorn harvest) in AD 806, 836, 935, 981, 1087 and 1108, but not in the years that follow. If oak mast had lessened by the early twelfth century, an alternative to acorns would have been needed. Cereals would have done that job well but at the expense of human effort, because cereals need tending whereas acorns come free.

8:3 One of the animals introduced during the second millennium AD, the mute swan

The twelfth century saw the arrival of two new animals to Ireland. These introductions mark the time when non-native species started to arrive in increasing numbers. The black rat (*Rattus rattus*) was introduced to Ireland in the mid-twelfth century and the earliest documentary record for rabbit (*Oryctolagus cuniculus*) is dated to AD 1191.

It was not until the fifteenth century, however, that the mute swan (*Cygnus olor*) and the hedgehog (*Erinaceus europeaus*) were introduced. The common frog (*Rana temporaria*) may have been brought in during Norman times or, yet again, during the seventeenth century. In 1697 frog spawn was introduced into the ditch at College Park, Dublin, after which frogs spread over the whole country.

THE EFFECTS OF THE BLACK DEATH

The pollen evidence shows grain grown throughout the second millennium but with breaks in the record for some places in the fourteenth century. Where the AD 1362 Öræfajökull tephra layer has been found in bogland, it pinpoints the decades immediately after the dreadful illness known as the Black Death, which has been attributed to an outbreak of bubonic plague that ravaged Asia and Europe in the mid-1340s. The cause of the plague remains contentious and there is little recorded about the effects of this terrible illness in Ireland. If the death rate in Ireland was similar to that in Great Britain and Europe, up to half of the population may have perished, having been already weakened by the poor harvests of 1314–1322. During this period of great duress there also occurred a major plague of cattle, culminating in 1315–1316. Large numbers of cattle died across Europe and it is supposed that in Ireland the disease was no less virulent. The loss of draught animals and of milking herds must have had a terrible effect on Irish agriculture.

At Garry Bog the AD 1362 tephra layer occurs in the peat at approximately the point where there is a break in the cereal record that lasted about 150 years and it is possible that the abandonment of cereal farming in the northeast of Ireland at that time may have been influenced by population change at the time of the Black Death. Perhaps there was a population collapse, with those people who remained abandoning the less fertile places and putting their efforts into the better land.

The tree-ring record for this period shows a lack of building timbers from most of the country between AD 1306 and 1350 and this dwindling could possibly have resulted from abandonment of managed woodland. There are dated timbers that post-date AD 1350 but at this time the cereal curve goes into decline for about 150 years. Could scraps of evidence point to a redistribution of population, with land close to bogland abandoned in favour of better land or land more easily worked?

THE MEDIEVAL WARM PERIOD AND THE LITTLE ICE AGE

In recent years, with increasing concerns about global climate change, there has been much research carried out on climate change since the end of the last Ice Age. The most recent work indicates that the climate of the second millennium AD may have experienced the greatest variation in temperature since the Ice Age ended. Various lines of evidence show that at the end of the first and then into the second millennia AD the climate of Europe first warmed – the Medieval Warm Period – and then entered several centuries of prolonged cold, from the fifteenth to the nineteenth centuries – the Little Ice Age. The dates of the onset of these periods remain highly contentious and, as yet, little is known about the impact these climate changes may have had on the Irish landscape. The results of many more studies will be needed before conclusions can be drawn. A few studies mark changes that started in the thirteenth and fourteenth centuries, which could conceivably be climate driven. At Dunfanaghy in County Donegal sand-dunes started to form about the same time as the onset of the cold period. Heath vegetation is uncommon in Ireland and largely confined to coastal sites but at inland Dunamase in County Laois the Great Heath started to form about 1500.

WOODLAND FROM AD 1000 TO 1700

The fate of the Irish woodlands at the end of the medieval period remains one of the most contentious in the history of the recent Irish landscape. There remains a strongly held belief, based almost exclusively on the historic narrative, that woodlands were very dense and extensive in Ireland until the late sixteenth century, after which time they had been almost exterminated. Some nineteenth century historians tended to over-emphasise the extent of the historic woodlands. For example, in her book on the woodlands of Tudor Ireland, Eileen McCracken states that it was believed that as late as the eighteenth century, '. . . there were remaining woodlands so dense that a squirrel could have crossed most of the country on the tops of the trees'. To try to assess the extent of the late medieval woods, one needs to be familiar with the story of the woodlands that emerges from history before comparing it with evidence from pollen analytical studies.

In accounts from the thirteenth to the seventeenth centuries, often written by the English military, it is said that the Irish rebels used the woodlands to mount ambushes and to provide hiding places. Felling the woods would, therefore, convert them to saleable timber whilst at the same time rid the countryside of refuges for skulking rebels. It was the commercial result of woodland clearance that was uppermost in the minds of those designing and implementing the scheme known as the Plantation of Ulster. The details of this early seventeenth century initiative to colonise or 'plant' much of the north of Ireland or Ulster with Scottish and northern English settlers is outside the scope of this volume, but having easy access to a ready supply of saleable timber was an enticement. Between about 1610 and 1640 portions of fine woodland, especially in the north, appear to have been ruined through uncontrolled felling.

Between 1610 and 1625 some people may have made money from carting timber to be sold out of woodland like the Great Wood of Glenconkeyne or Killetra in the Lower Bann valley to Coleraine. It cost as much as 10s. 6d. per cart-load to carry the wood, an exorbitant price when compared with the cost for general cartage of 1s. 6d. Much of the felled wood never left the woodlands. There are descriptions of ruinous wastage in the Great Wood of Glenconkeyne in County Londonderry where newly felled trees were left to rot where they lay. Between 1600 and 1650 the woods of the Lower Bann were so stripped of trees that once wooded places were opened but choked with tree stumps. These too were grubbed out to leave the rolling countryside along the banks of the Lower Bann River free of woodland and scrub. Today the little hills have only the trees of the hedgerows and nothing remains to show where great woodlands once stood. Further afield the woodlands of County Donegal, never as prolific as those to the east, were also reduced to patchy scrub.

8:4 It is possible that the fine woodlands of the Lower Bann Valley were reduced to patchy scrub

The only place substantial woods remained was on the mountain slopes running down to Mulroy Bay. There, as in numerous places throughout the country, woods remained mostly in areas that were awkward to access and where there was more trouble than gain to be had in cutting trees for timber or fuel. Written evidence about alder woodland on the banks of some of the slow-flowing midland rivers and the felling of this in the seventeenth century indicates that its loss may have increased the extent of the herbaceous, species-rich water-meadows or 'callows', as they are known in Ireland.

There were laws passed to keep the big northern woods for naval use, although not for ship-building timber. Oak wood was the source of wood for the staves from which casks were made and wood was also processed for charcoal, a ready supply of which was essential to the workings of the small ironworks and glassworks that sprang up in considerable numbers. Ironworks or 'bloomeries', as they were known, are recorded throughout the country and they were expensive to set up, each costing £1,000–£3,000. It took about 1.25 tons of charcoal to smelt 2.5 tons of ore and the iron produced varied in value, depending on the port at which it was sold: for example, locally produced iron fetched £11 per ton at Ballyshannon but £17 per ton if sold in London. The unceasing demand for abundant charcoal placed great pressure on the woods and once they were destroyed the iron industry ceased.

Little is known about woodland management practices in Ireland at the end of the medieval period. Some useful information about earlier woodmanship can be gleaned from monastic records and the recorded lives of saints, from which sources it is plain that coppiced woods were considered a valuable resource within a highly structured agricultural landscape. This practice must have continued. In the State Papers from the early seventeenth century there is occasional mention of hazel coppices in the north of Londonderry and of an extensive coppice wood between Dunalong and Lifford in County Donegal but little is recorded about coppiced woods anywhere else in the country.

A better indication of woodland management emerges from the large numbers of dated timber that comprise the Irish oak dendrochronological archive. For the mid-fourteenth century in particular tree dates often start at 1350, possibly marking a resurgence of woodland in the wake of population collapse after the Black Death. Had timber been obtained from unmanaged woodland, it could be expected that the start date for the building timbers from those times would have spanned a period of time.

Very little has been written about the country-wide extermination of hazel scrub that followed the depletion of the larger woodlands in the middle years of the seventeenth century. In contrast to the scant documentary evidence for yet another alteration of the woodlands, the rapid and extensive loss of scrub is clearly marked in pollen profiles. Hazel is a most prolific producer of pollen and where large falls

8:5 Today turf is still cut using the spade

in pollen percentage values are noted, they mark a major loss of this once common shrub. Various means must have been used to get rid of the hazel scrub, including the hard work of grubbing out the plants. The great reduction in hazel is broadly coincident with the end of the old Gaelic order in the seventeenth century and with it the loss of a farming lifestyle that had prevailed for centuries, possibly millennia.

With the loss of so much wood and scrubland in the seventeenth century, the dirth of fuel was met partly by the cutting of peat or 'turf', as it is known when intended for burning. Turf had been cut hundreds of years before this time, as the first millennium Brehon Laws testify, but not on the enlarged scale of the eighteenth and nineteenth centuries. As the need for turf grew, it was the edges of the lowland bogs that suffered first, so much so that today there is no bogland remaining with an intact 'lagg' or edge. The peat in the uplands and also in the interdrumlin hollows in the north of the country provided ready sources of fuel but it was hard-won, as it had to be dug, dried, stacked and carried. Cutting turf is a job to be done in good weather with a drying wind.

8:6 Access to turf cutting was restricted in some places

NOTICE.

Any Person found Cutting Turf, Raising Bog Fir, or committing any Waste or Injury to or upon my Property in Ennishowen, after this Notice, will be Prosecuted and Punished according to Law.

All former liberty is withdrawn from this date.

Redcastle, April, 1877.

J. M. DOHERTY.

When the pollen and written records for Irish woodland in the seventeenth century are joined, a truer picture of the fortunes of the woodlands emerges. At no time in the last two thousand years have there been woodlands in Ireland that matched in extent and density those of ancient times. Doubtless some of the largest woodlands, such as those in Killarney, were old and biologically diverse. Studies of the fossilised remains of woodland beetle species support the view that good quality, possibly managed, woodland survived. Where woods were wiped out, however (and this was the general trend), many species of birds and insects must have been made locally extinct.

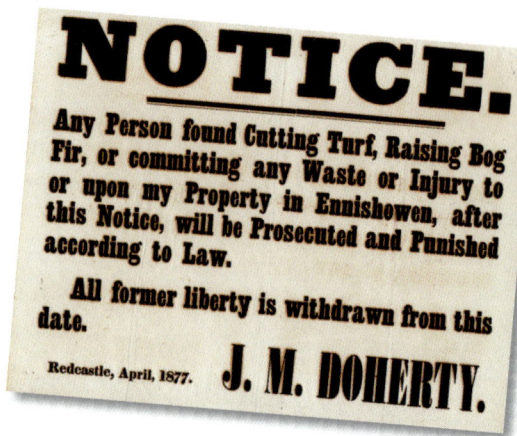

CHANGES THAT BEGAN IN THE 1600s

The loss of the hazel scrub was not the only change in land management practices that took place. From approximately the middle of the seventeenth century, the changes which the Irish landscape underwent can hardly be overemphasised. Many changes occurred in the years following the Plantation of Ulster during the first two decades of the seventeenth century and the 1641 Rebellion. Perhaps the most marked change in agriculture at that time was the introduction of the potato (*Solanum tuberosum*), first as a garden and then a field crop. The history of the introduction of the potato to Europe is not well recorded and, in keeping with this general pattern, there is little in the historical record to mark the introduction of potatoes to Ireland. It is known, however, that as early as the 1620s potatoes were being grown on a small scale on the Ards Peninsula in County Down. It is particularly unfortunate that the landscape's record of the potato is almost as silent as the written word on the topic. Potatoes have left almost no trace in the pollen record, partly because the pollen does not fossilise well.

Potatoes were introduced from southern America, having originated in the Andes, and they were unlike any crop that had been grown in Europe before. Until the introduction of the potato all major crop yields had come from the aerial parts of the plant in cultivation. Not so with the potato. The edible tubers grow underground and need careful and frequent tillage to give a good harvest. In contrast to most other root crops that eventually shrivel in storage, potatoes remain in good condition months after harvesting. The potato has been described as 'the best all-round bundle of nutrition known'. Almost 80 per cent of a potato tuber's mass is water, with the remaining solid material made up mostly of carbohydrate

8:7 Potato plants thrive in Ireland's mild, moist climate

and protein, along with vitamins and essential trace elements. The protein content of potato is low when compared to cereal grains but the tuber is rich in essential amino acids. A diet of potatoes and milk is nutritionally complete and able to sustain an adult for months if two to three kilos of potatoes per day are eaten. Many people lived on little other than boiled potatoes, buttermilk and a little salt to flavour the meal. Any other foods were an occasional luxury.

During the heyday of the potato, from about 1750 to 1810, the Irish landscape was briefly dominated by arable agriculture – for the only time in Irish farming's long history. Potatoes were often grown on raised ridges known as 'lazy' beds and it is within the author's experience that digging a lazy bed is not an exercise for the lazy. Where grazing land went over to tillage, disturbance increased and weeds flourished on the newly opened soil. On the sides of the modern ridges (and presumably also those in the past) annual chickweed seeds germinate under the shade of the potato foliage and in the troughs between the ridges goosefoot and knotgrass grow strongly in the slightly moister soil.

The spade was the universal tillage tool and it was used throughout the country to change the function of thousands of acres of land from meadow and pasture to open, friable ground. Some of the soil from the loosened surfaces was washed into lakes and at Lough Neagh the lake sediments show accelerated mineral in-wash, starting about AD 1700. The increased demands placed on the land by arable farming damaged soil structure and nutrient levels. The land under potatoes increased and, in the north of the country, flax was also more widely grown, its cultivation further depleting the soil of its nutrients as it is a 'hungry' crop. By the mid-eighteenth century the demand for wool, beef and butter had increased enormously and the massive exports of cattle accounted for further reductions in the levels of calcium and phosphorus in soils.

NEW TREES ON THE LANDSCAPE

By the end of the seventeenth century wood was in great demand, commercially and domestically, and it was at this time that non-native tree species were first introduced. After a gap of almost four thousand years Scots pine was reintroduced. It may have been brought back both as a garden ornamental and as a woodland crop. In spite of accounts describing the first new pine seedlings being brought in by Cromwellian troops in the mid-seventeenth century, the tree was introduced by the end of the seventeenth century, later to be joined by other conifer species from distant places. Saplings of larch (*Larix europea*) and spruce (*Picea abies*) were registered when planted so that the owner could claim a bounty when the trees were later cut.

In 1682 it is recorded that, at his estate in County Longford, Lord Viscount Granard planted a range of non-native tree species that included pine (*Pinus* species), cedar (*Cedrus* species), lime, beech and the famous 'Platanus tree'.

It is thought that this tree was unique in his time, in that it grew nowhere else in Ireland, and that he had planted the first sycamore (*Acer pseudoplantanus*), a tree today found growing all over the Irish lowlands. The pollen record does not abound with pollen grains of the new alien species but occasional pollen grains of sycamore appear in pollen profiles dated to the eighteenth and nineteenth centuries. At about that time horse chestnut (*Aesculus hippocastanum*), a native of mid-Europe, as well

8:8 The famous Platanus tree may have been a sycamore

as lime and beech, natives of the woodlands of Great Britain from earliest times, were also introduced. It was not only new types of trees that were introduced during this period. Until then native salmon and trout were the major freshwater species but with the introduction of gudgeon (*Gobio gobio*), bream (*Abramis bramis*) and rudd (*Scardinius erythrophthalmus*) the 'coarse fish' soon became established.

THE RISE OF THE HEDGE IN THE EIGHTEENTH CENTURY

The Irish agricultural landscape has been divided by walls since Neolithic times with the extensive system of low stone walls at Céide in County Mayo. Walls continued to function as boundary markers throughout Irish prehistory but the history of the hedge in Ireland is barely recorded. The Brehon Laws comment on hedges being used in management of the farmed landscape, and in the *Calendar of State Papers*, Ireland, from the sixteenth century there are occasional comments on hedges within woodland that are blamed for making soldierly progress difficult. Such hedges were described as 'plaited'. Through partial cutting and bending, known as 'laying', the branches of shrubs that make up a hedge can be knitted together to strengthen and thicken the whole structure; from these historic accounts it is surmised that hedge management techniques were known and practised.

By the early eighteenth century hedges had an enlarging part to play in land management and division. In County Armagh it is recorded that hedges had been planted and other accounts describe a preference for hawthorn or quickthorn. Hawthorn, a member of the rose family, makes an ideal hedge shrub as it is thorny, grows quickly and can be 'laid'. So great was the need for young thorn plants that large numbers of 'quicks', as they were known, were imported from England. Thus the scrub-free landscape enclosed by hedges that is characteristic of the modern Irish lowlands has its origins in the early eighteenth century although walls continued to be used in the uplands, where hedges do not thrive, and on stony soils.

Hedges had a further function. They acted as a renewable source of wood and into them farmers were encouraged to plant larger, rapidly growing trees like ash and sycamore. When an ash 'plant', as it is often called, is cut back, branches spring readily from the basal 'stool' providing light timber, as the wood of ash is smooth and ideal for shafts and handles of tools as well as for the essential hurling sticks. Sycamore was useful in butter-making as vessels made from its wood, which, having no taste, did not taint the milk. In overview the 100-year period from about 1650 to 1750 saw major alteration to the face of the lowlands, where hedges marked potato, flax and cereal fields as well as pasture for the farm animals.

THE RISE AND DEMISE OF THE POTATO

From 1687 to 1791 the population of Ireland increased from 2.16 million to 4.75 million, with a further increase to 8.15 million people by 1841. After 1740 and by the end of the Napoleonic wars in 1815 there were 4.7 million people in Ireland, of whom 3.3 million had nothing else to eat but potatoes. Irish farming had become a 'potato monoculture', leaving the entire agricultural system vulnerable to any change in growing conditions that might adversely affect potatoes. The decades between 1815 and 1845 saw a series of poor summers in which potatoes did badly and it was the poorest people who suffered most as the few potatoes they were able to harvest were their sole source of food.

The full story of the Great Hunger or Irish Potato Famine of the mid-1840s is beyond the scope of this volume, but it was the potato blight fungus (*Phytophthora infestans*) that was the cause of the devastation because it killed the growing plant.

Once infected, the plant rapidly sickens and sags into blackened, foul-smelling decay and, to make matters worse, the wet weather of the famine years favoured the spread of the fungal spores. Acres of once-green foliage turned black within days of infection and the stench of rotting vegetation pervaded the countryside. In places where the crop had yielded tubers, these were collected and stored, but to no avail as they too succumbed to the rot and any that seemed fit for use as 'seed potatoes' carried the infection into the following year. It had been predicted that the harvest for 1845, the first year of the Famine, would be 15 million tons from 2.5 million acres of land. In reality there was no potato crop that year nor in the two years that followed.

The origin and spread of potato blight remains contentious but recent research has pointed to the route taken by the pathogenic fungus, at that time unknown to science, as it entered and spread in Europe. Historical records and modern genetic studies point to the country of origin of the fungus as Mexico, from where, hundreds of years earlier, it had migrated to the countries of the southern Andes. In 1841–42 there was a further migration of the fungus from South America to the United States. It may have been 'seed' potatoes that carried the fungal pathogen from South America, the United States, or both, to Belgium where 'seed' potato stock had dwindled. Locally grown potatoes had suffered from viral infection and tuber dry rot caused by *Fusarium*, with which potato blight was first confused.

Potato blight may have been present in Europe in 1844 and locally as early as 1843–42 but not in 1841. In the spring of 1845 the fungus spread out of Belgium and about ten weeks later it was detected in France, the Netherlands, and the British Isles. The first signs of the disease in Ireland were noted by David Moore, the Curator of the Botanic Gardens at Glasnevin, on 20 August 1845 and it was seen in potato fields immediately thereafter. In September of that year its effects on the

8:9 Foliage of a potato plant 'blighted' by infection of the potato blight fungus

Irish potato crop had become devastating and by mid-October all counties had become affected. At the time of the first ravages of the disease there was no agreement on its cause, with poor weather conditions and even the electrical state of the air being blamed as much as a fungal plague.

THE POTATO IN THE UPLANDS

By the mid-1840s, to meet the needs for food of the expanding population, upland areas previously considered unfit for crops had been taken over by the poorest people who tried to grow potatoes in the wet, peaty soils. In the hills of the west of Ireland and in upland areas far from the coast people worked at digging lazy beds along natural drainage lines. They planted their seed potatoes and waited, in hope of a harvest that would feed them from midsummer to the spring of the following year. The outlines of the lazy beds are still visible on hillsides throughout the country and it is often said that the ridges have lain abandoned since the time of the Hunger.

Did the potatoes fail to grow or did their meagre harvest rot because of the all-pervading presence of the blight? Some long-abandoned ridges testify to the latter, a failure of the harvest. Examination of abandoned ridges on Slieve Gullion in County Armagh by the author showed that they had never been dug to harvest a crop.

The largely failed effort to grow potatoes in the uplands was the second and last time that the hillsides of Ireland were brought under tillage. More than 3,000 years before, during the Bronze Age, the uplands had first been cleared of woodland, opening the soils for crops and pasture. For a brief period then the farmers had

8:10 Abandoned lazy beds on Clare Island off the coast of County Mayo continue to show their ridge and furrow structure

successfully harvested crops and kept cattle. By the mid-nineteenth century, however, farming was doomed to failure because the upland soils had long given way to nutrient-poor blanket peat. In the spring and summer of the years leading up to the mid-1840s the colour of the hillsides changed as patches of brown peat were dug over in preparation for lazy bed construction. With their abandonment, however, the native vegetation grew back, bringing a summer flush of pink and purple as the heathers flowered again.

In 1845, 1846 and 1847 it is estimated that one million people died as a direct consequence of starvation and a further million from dysentery and cholera. Those who could gather together enough money for a passage to England or the New World left in the hope of a better life. The old and destitute were left to carry on as best they could. Farms that had been worked by succeeding generations were left derelict. Throughout the country the little clusters of homes known as 'clackans' were abandoned and, as the years went by and the houses tumbled, the roofless, low walls sheltered sheep from hard winter weather. It has been said that to this day the Irish landscape remains blighted by the ravages of a famine that robbed the country of its population.

THE ABANDONED LANDSCAPE

In the years that followed the Famine woodland began to regenerate in some places. There are patches of woodland scattered throughout the country now that are denser than they were in the mid-nineteenth century. Tree-ring dating has proved that many of the oak trees alive today germinated in the years after the Famine and even big trees with substantial trunks are often only about 150 years old. In spite of the ravages potatoes were not abandoned as a crop. The development of fungicides like 'Bordeaux' and 'Burgundy' mixtures let small farmers keep potato blight under control by spraying the leaves against invasion by the fungus, but potatoes were never again grown on the huge scale that characterised Irish agriculture in the early decades of the nineteenth century.

CHAPTER 9

RECENT TO FUTURE LANDSCAPES

100 years ago to the present

KEY ISSUES:

bogland exploitation;

plant and animal alien species;

climate change.

If one looks closely at the surface layer of much uncut, lowland raised bog peat, it is often darker in colour and drier than the peat below. Although bog surfaces are soaked regularly by the rain, on the cut and drained bogs more water is lost through seepage than gained through rainfall. The result is that the top peat dries, the air enters and the peat darkens as it oxidises. On numerous bogs the mossy vegetation has given way to more drought-tolerant heathers and bog myrtle (*Myrica gale*) and some remnants of lowland raised bog have become so dry that *Sphagnum* peat no longer forms.

To varying extents these changes have adversely affected the hydrology of the remaining uncut bogland throughout the country. Considering that Irish lowland peatland, known by the term 'oceanic raised mire', represent a unique ecosystem, its continuing degradation is of international consequence. The damage being done to boglands is not unique to that ecosystem; others are also threatened, often as the result of the actions of people.

The earlier chapters of this work have told the story of how the Irish environment has changed over centuries and millennia but this final chapter will not follow strictly that earlier pattern. This is because the landscape of Ireland has undergone so many changes over the last 100 years that no single chapter could do the story justice. For example, the farmed landscape has been altered greatly by mechanisation and the introduction of new crops and farming techniques. Woodlands and species-rich grasslands, extensive a century ago, have shrunk enormously in extent, whilst areas of the uplands altered fundamentally once planted with non-native conifers. In the 1950s and 1960s, throughout the uplands, hundreds of hectares of conifer plantation came to dominate previously open places where few trees had grown for over 4,000 years.

There are many detailed accounts of the manner in which the Irish landscape has altered, especially since the end of the First World War, told fully and authoritatively by other authors.

The aim of this final chapter is to provide information that may inform the measures needed to conserve places that retain some of the character of times past. Many of the comments in this chapter draw upon the author's observations over the last sixty years and mostly in the north of the country. In spite of some inevitable

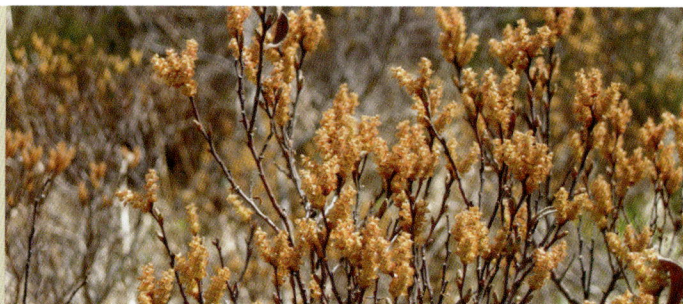

9:1 Bog myrtle in flower. The small shrub does well on dried bog surfaces.

Facing page:
9:2 Conifer woodland planted during the second half of the twentieth century

geographical imbalance in the account, it is hoped that the information that follows will contribute to the debate on ways to safeguard the fragments of the older landscape that remain. I regret to record that modern conservation measures are not always informed by past environmental studies. In Ireland, as indeed in many other countries, there are numerous threats to the well-being of long-established native species and ecosystems. If individuals as well as local and national bodies have little access to information about their local landscape, past and present, they may struggle to grasp that wider environmental change may well have local impact.

The preceding chapters of this book drew heavily on the evidence for past landscapes revealed through studies of peat, therefore the status of bogland in the first decade of the third millennium AD is worthy of comment. The 1,170,000 hectares of lowland peat that remain are a fragment of the 134,000,000 hectares that once dominated 16 per cent of the Irish land surface. Seventy-eight per cent of lowland bogland has been cut and many once-extensive raised bogs have been drained, fertilised and converted to pasture. In the years following the end of the Second World War in 1945 huge tracts of lowland bog in the midlands have been harvested as fuel for electric power stations. It has been estimated that 46 per cent of blanket peat has been cut in the past, leaving the intact blanket bogs that remain representing 8 per cent of the world's blanket peat resource, therefore any further reduction in the extent and quality of Irish blanket peat, upland and lowland, is of global significance. In recent times there has been an upsurge in domestic peat harvesting. Peat for the family fire or for the small-scale retail trade is cut mechanically. Whether mechanically cut for the power station or the family fire, the bog surface is denuded of vegetation and open to erosion by the weather.

These figures prove that the big bogs, the oldest of which began their development 11,500 years ago, have dwindled greatly in the recent past and it is not only the peatlands that have suffered in this way. Deciduous woodland, species-rich grassland and other native ecosystems have also diminished in size and biological diversity, with many reduced to scattered scraps. All of these threatened environments need protection from further encroachment and loss.

The state of many upland areas provides a good example of habitat fragmentation and degradation primarily through erosion, which may have a lengthy history in some parts of the country. In the Mourne Mountains, for example, some areas of blanket peat have been eroding, possibly intermittently, for about 4,000 years. The cause of erosion in the blanket-peat-covered highlands is not well understood and it is unclear how significant the contribution from deteriorating climate or human exploitation may be. Erosion may be inevitable in areas where sub-peat pipes drain bog pools and destroy the stability of large masses of soft, wet, organic material perched on a slope. The detrimental impact of erosion in upland areas that include reservoirs can be surmised readily.

Throughout the country there are areas of damaged peatlands whose future use requires research and planning. There are no straightforward solutions that will easily resolve the difficulty of what to do with damaged upland areas. There is little point in planting forestry on degraded upland peat where it is expensive to maintain. One of the more novel uses of former commercially exploited peatland in the lowlands has been to flood it for use as a wildlife refuge, like the successful Lough Boora Wetlands in County Offaly.

Would that cut-over boglands would cease to be used as a dumping ground, for they are often remarkably rich in species of plants and animals. I have regularly seen dragonflies hunt over the pools of water at the bottom of abandoned turf cuttings and have startled hares when walking across a bog. Cut-over bog is often home to larks, which are becoming uncommon and whose singing on a warm summer's day, high above the heather and cotton grass, is unrivalled in beauty. On a more positive note, it is heartening that many of the best examples of remaining unexploited bogland in the Republic of Ireland are under the protection of the Irish Peatland Conservation Council and their efforts to educate people about the wide-ranging value of bogland cannot be praised highly enough.

If bogland has been thought of by some as having little value then the opposite must be the case for woodland. Previous chapters have described those ancient times when Ireland was heavily afforested. Of the 5 per cent of countryside currently under woodland only 1 per cent is made up of mixed native species, the remaining 4 per cent being composed of non-native conifers, growing mostly in large stands of single species such as Sitka spruce (*Picea sitchensis*) or lodge-pole pine (*Pinus contorta*).

Internationally, Ireland is recognised as having the least tree cover in Europe but palaeoenvironmental research has demonstrated that our relatively treeless landscape has a history lasting at least 4,000 years. Is it perverse, therefore, that in a country with a tradition of validating modern practices by appealing to the past, there is renewed insistence on native planting trees? Where areas of old woodland can, however, be extended and strengthened by new planting, the future of the remnant can be secured and its ecological diversity enhanced.

It is worthy of note that the term 'ancient' is used of woodland that has been present for a few hundred years. When compared to the first development and duration of the former great forests, the term 'ancient' is exaggerated but the richness of the oldest woodlands in plant and animal species makes them worthy of conservation. The oak woodlands of Killarney are amongst the oldest woodlands that remain in the country and, although still large, they are less extensive than once they were.

Attempting to identify seemingly old woodland not marked on early nineteenth-century maps may present difficulties – just because woodland looks old does not mean that it is old. Consider the case of the ash woodland at Hanging Rock Nature

9:3 Throughout the country species-rich grazing land has given way to silage production

9:4 Peat cut mechanically for domestic use

9:5 Huge swathes of the midland bogs have been denuded of vegetation, leaving the bog surface open to further erosion

Above: **9:6** Lough Boora wetlands support increasing numbers of plant and animal species

Below: **9:7** Tenacious weeds will thrive even when faced with harsh conditions

Reserve in County Fermanagh. Many species of plants grow in the wood at Hanging Rock and the rare pine marten has been seen there. The appearance of the place gives it an old and well-established air but there is a mid-nineteenth century engraving of the Hanging Rock showing the area almost treeless. The wood may not be 'ancient' but it is rich in native species and worthy of conservation, especially as extensive ash woodlands are not common anywhere in the country.

People respond in various ways to the conifer woodlands that have become a feature of the Irish landscape since the late eighteenth century. Forest parks are used for all manner of leisure activities and the timber grown makes a valuable contribution to the national economy but it is the appearance of these woodlands that displeases the eyes of some because they look 'unnatural'. Some people dislike the sharp edges of the conifer plantations. Do they dislike equally the defined edges of the fields of golden, ripe cereal? The dull green of the spruce and larch appears ugly to some but there was a time when the uplands held large tracts of pine wood and it was just as sombre in colour as the new woodlands, admittedly without the defined woodland edge that is characteristic of modern plantations. Would knowing that the old pine woodlands were dark green influence how some people react to the most recent type of woodland that grows in the Irish uplands?

Grasslands remain extensive throughout Ireland. Some are very recent, such as the fields managed for silage production throughout the country. Others, perhaps never ploughed as in the Curragh of County Kildare, have increased in wealth of species over the hundreds of years of their existence. In addition to the higher plants the Curragh has an unrivalled diversity of fungi. Grasslands that have been tilled or fertilised at some time in the past can return to former species richness if the seeds buried in the soil are left to grow. As the flora of grassland becomes increasingly diverse, insects, spiders and birds rise accordingly in species and abundance. It is heartening that farmers in designated areas are paid to maintain their grasslands using older management practices, in well-researched efforts to conserve habitats for rare plant and animal species. It is also interesting to note that about 52 per cent of farmed land is non-arable and of the 48 per cent that is arable land, the amount under cultivation seldom exceeds 10 per cent, making practices that safeguard the tilled landscape and its unique plants and animals also worthy of support.

The weedy waste ground has received limited recognition from conservationists. Whilst looking untidy these places retain something of the character of the Irish landscape before the last Ice Age ended. Many of the wild plants that grew throughout the country in those ancient times have since been relegated by people to the status of weeds. Their former place in the development of our native landscape remains unacknowledged and it is true that today these plants thrive where people have opened up the soil. The weedy places should not in every case be swept away, considered as nothing but a nuisance.

9:8 *Rhododendron* has invaded native woodland and other ecosystems such as blanket peat

They do not have the majesty of ancient woodland but they are by far the oldest ecosystems that remain and they deserve proper recognition.

The native kestrel (*Falco tinnunculus*) once depended on the small mammals that abounded but today the bird may also eat the young of rats that came to this country in the wake of people. We can no longer separate the fortunes of the native from that of the newcomer, since the activities of people over the last 6,000 years have muddled the native and newcomer relationship. Perhaps we must learn to work with what we have, rather than harking back to some poorly defined notion of purity that we know to be supported only weakly by the incomplete Irish native record.

Whilst recognising that the line between native and non-native is blurred and has been for a long time, one cannot ignore the impact of alien species on the Irish landscape – and the aggressive aliens in particular. Ireland is full of aliens! Most of the alien plants arrived about three hundred years ago, the animals more recently brought in from abroad by wealthy people to beautify their estates. Some of the plants in particular did unexpectedly well under Irish conditions. *Rhododendron* is a good example of an ornamental plant that escaped from the 'big house' garden to establish itself in woodland and on cut-over bog. The shrub, which bears familiar

9:9 The flower of another alien plant, the giant hogweed, towers above a thicket of Japanese knotweed foliage

and spectacular purple flowers in spring, comes from an area spanning Turkey and Bulgaria as well as Lebanon and Syria and into western Asia. The Worldwide Fund for Nature categorises rhododendron as a serious and widespread invader and the plant that does so well in Ireland may be a hybrid of *Rhododendron ponticum* and *Rhododendron catawbiense*, a North American native.

The hybrid spreads by seed and it can grow on many types of soils as well as showing a preference for cut-over bog. Once established in Irish woodland, it casts shade and drops leaf litter. The rare mosses and liverworts of the native woodlands of Killarney are greatly threatened by the spread of rhododendron, which has few natural predators even within its native range. It shrugs off chemical weed-killers and, to make matters worse, appears to flourish in places where efforts have been made to clear it by hand.

More recently the rhododendron has been joined by other aggressive aliens. These include the Japanese knotweed (*Reynoutria japonica*) that has invaded woodland edge throughout the British Isles and which is so vigorous that it has been seen growing through concrete road surfaces in city centres. A native of Japan, north China, Taiwan and Korea, it was introduced to Europe in the mid-nineteenth century as an ornamental and fodder plant. The bamboo-like stems grow to 2–3 m high. The plant flowers late in the summer but does not spread by seeds because all the plants in Europe are female and genetically identical. The minute fragments of its underground stems grow into new plants and its growth habit retards re-establishment of native plants from the soil seed-bank. The knotweed's woody, dead stems stand all winter and the new shoots in spring emerge amongst the previous year's dead growth to form dense thickets, through which native seedlings cannot emerge. It comes as no surprise that the plant has proven very expensive to control using chemical weed-killer, mostly because the underground stems let the plant regenerate even after the top growth has been killed. At a site in England an area infested with Japanese knotweed measuring 30 m by 30 m cost developers £52,785 to eradicate.

The waterways of the country are rich in native species of plant and animal life but they have come under grave threat since about 1990, just after the Wildlife (Northern Ireland) Order of 1985 made it an offence to deliberately introduce invasive alien species. I have observed some extraordinary plant successions based entirely on alien species whose interactions have drastically and detrimentally affected former assemblages of native aquatic species. The example that follows is given so that it can be contrasted with the account of the steady but not dramatically rapid spread of the native plants provided in an earlier chapter of this text. These modern successions of alien plants are probably more rapid than any that occurred in the past.

There follows an extraordinary example of a plant succession involving aggressive aquatic alien species growing at a site in County Down and which took

9:10 A pond in County Down in which alien plant species have become rife. Australian stonecrop grows in the centre of the pond and floating pennywort around the edge.

only about six years to complete. Irresponsible dumping of the contents of a freshwater aquarium has been blamed for the introduction of these plants to a site comprising a number of deep ponds. In the late 1990s the New Zealand pygmy-weed or Australian swamp stonecrop (*Crassula helmsii*) had formed a thick mat just below the surface of largest pond. Soon after the arrival of the pygmy-weed, the tropical water fern (*Azolla filiculoides*) was seen, followed by floating pennywort (*Hydrocotyle ranunculoides*).

In the first couple of years the pygmy-weed grew profusely all over the surface of the largest pond. A couple of warm summers followed that suited the *Azolla* better and when it dominated the pond surface it cast such dense shade that the pygmy-weed beneath died out.

Harsh winter weather probably killed off most of the *Azolla* but not the floating pennywort. From a small patch in the corner of the largest pond, where it had held out against the invasion by the two previous plants' rapid growth, the floating pennywort spread all round that pond's edge, then smothered the surface of the smaller pond nearby and began to grow up the trunks of trees at the water's edge. The floating leaves of the plant remain green even in winter and in summer the vegetation mat is strong enough to support the weight of numerous mallard ducks

9:11 The red-brown leaves of the tropical water fern *Azolla* can smother the water surface in a few weeks during summer

and a swan family. The tragic fate of the former native water plants must be presumed.

This tale is told to caution against any assumption that non-native plants will not threaten the large native ecosystems. The story of the invasion by the zebra mussel (*Dreissena polymorpha*) along the full length of the Shannon Waterway in about fifteen years and the damage done by the mollusc to the native aquatic plants and animals is well documented.

9:12 Zebra mussels are now so common in the waters of Lough Derg that millions of their shells are cast onto the shores after winter storms

9:13 The grey squirrel

None of the alien land animals, not even the despised grey squirrel (*Sciurus carolinensis*) has had such a detrimental effect on native species or ecosystems as that caused by the relentless spread of the zebra mussel. An internet search will provide the detailed information the reader may need.

To add to the problem now approaching national proportions, the extermination of alien species has proven expensive. In 2002 the cost of removing New Zealand pygmy-weed from ponds in the New Forest was estimated at £60,000–£110,000. In Great Britain an experiment to control the spread of the water fern using the leaf-eating larvae of the *Azolla* weevil (*Stenopelmus rufinasus*), a native of North America, may offer a means of control. Perhaps over time some invasive species may reach equilibrium with some of the natives and this may lead to better efforts to eliminate those that remain stubbornly pernicious.

It has become clear that the impact of human activity and the spread of alien species have impeded conservation of native ecosystems, but what of the impact of future climate change? The climate has played a huge part in the development of the Irish landscape since the end of the last Ice Age, as the first chapters of this text made clear. An astonishing fact to emerge from past environmental studies has been the rapidity with which the past global climate has changed. When the last Ice Age ended, the climate deteriorated catastrophically in just a few decades – could something similar happen again? Perhaps the answer to that wide-ranging question is 'yes', but the nature of that change, the rate of change, and the potential impact of global climate alteration is still poorly understood. Experts agree that the climate is changing, partly in response to human impact on the environment over the last two hundred years, with the rate of change accelerating during the last century. American politicians have denied vehemently that human-induced global warming is happening, and yet their top scientists, funded by public money, regularly publish their findings in the international scientific literature. American palaeo-climatologists lead the world in the studies of past climate change and they report findings that support studies of global warming accelerated by human impact. They also study the long-term natural climate cycles and attempt a synthesis of the findings from numerous studies. How any such changes will affect our native ecosystems is not, as yet, well understood but one may surmise that the detrimental effects will worsen if brought about by alien species suited to the warm conditions forecast for the next half-century.

The following example is of a possible interaction between a warming climate and the peatlands, in which we continue to be relatively rich. There are large areas of peatland to be found in Canada, Siberia and some tropical countries, therefore any changes brought about by transitions in peatland are not limited to the British Isles. The slow decomposition of peat produces carbon dioxide and methane, two of the most powerful of the greenhouse gases. Rising temperatures associated with a growing greenhouse effect may speed up decomposition of peat and the resultant increase in atmospheric concentration of these greenhouse gases released could, in turn, further accelerate the global warming process.

So what can or should be done to restore or 're-wild' places that retain some of their original character and what part does the non-specialist have to play in any action taken? There has been a global debate on the ethics and the realities of attempting to return ecosystems to the perceived conditions of the past and there are many different interpretations of what is meant by 'restoration'. The reader should be aware that this wide-ranging debate is ongoing and vigorous. Restoration, preservation, conservation, stabilisation – call it what you will – of ecosystems involves much more than the application of scientific findings and technological advances. There are cultural issues that also deserve attention. For example, the complex interplay, some say conflict, between modern bogland conservation management techniques and the age-old practices of cutting turf from the bog for the family fire, or indeed the power station. There are no straightforward policies to satisfy everyone concerned with bogland, or indeed any other ecosystem management. In the face of these many difficulties, we acknowledge our responsibility to safeguard from further degradation the fragments of the wild places that remain.

During my academic life I have gained recognition as an expert in Irish environmental change since the end of the last Ice Age. I have broadcast on these topics on radio and television and taken great pleasure in writing texts for people who care about what happens to the Irish landscape. As is the case for most people, I do not have great influence with the government organisations charged with the responsibility of developing management strategies or making decisions. From time to time I am asked to inform these people and the advice I provide is as relevant to the individual non-specialist as it is to the people in authority.

What can the individual or groups of individuals do? School and community groups have a part to play in safeguarding local habitats. There are initiatives that give support to local groups and Ireland has responsibilities to safeguard its native species at European Union level. All should be aware that the need to preserve and conserve is urgent and accept that the best of the ecosystems that remain may be little more than fragments. The countryside still holds many scraps of woodland, grassland and scrub that scientific studies show retain much of their earlier character. At a national level it would also be enormously beneficial to the well-being of the less well-known ecosystems if their importance could be recognised as equal to that of the bogs and the woodlands.

There is nothing that can excuse the destruction of a salt marsh resulting from a thoughtless action taken to clean a plough. Such a case was reported in the recent past as having happened at a salt marsh on the shores of Strangford Lough. The salt marsh may not appear as eye-catching as the woodland on the nearby drumlin slopes but it is every bit as important an ecosystem. Individuals who have the privilege of owning such places can make it their personal responsibility to protect them. Many people do care, and an enquiry to a governmental agency may reveal sources of funding for maintenance, however there are people who place lesser value on the wild places, especially if their conservation hampers economic development. One may find oneself challenged by this alternative way of seeing things, therefore all of us, of whichever persuasion, need to know why we take the stance we do. We cannot force consensus on the management of our landscape, however precious some of us believe it to be.

In recent years, interest in wildlife issues has become 'big business'. Eco-tourism is an international money-maker. Through the media we hear much, for example, about gardening responsibly to encourage wildlife to thrive. We are discouraged from using peat-based composts as well as encouraged to let the slugs have their share of our garden produce, all in the interests of the common, greater good. Well-intentioned advice about letting small areas of the garden revert to nettles or some similar wild plant to encourage native insects probably does no harm but there is little in the way of quantitative research to prove that such practices actually encourage birds and insects back to places from whence they have been driven. Availability of food stuff is not always the limiting factor.

Perhaps we need to view these pieces of advice differently and accept that unless we are familiar with nettles and other less-loved but vital plants – ragwort springs to mind – then we are less likely to appreciate that they have vital role in the functioning of complex ecosystems. By all means grow primroses along the garden path but use the experience of seeing them close at hand to appreciate the beauty and importance of an extensive hedgerow in its springtime beauty. Grasp that, rather than be deluded that the plants in the garden are automatically part of a continuous woodland web, which is rarely the case as the fragments of native places that remain are often isolated one from the other.

Are there places where wildlife can thrive uninterrupted across the entire country? There are. The wildlife corridors formed by miles of farm and roadside hedges are seen frequently, and often by people who would not have the regular opportunity to visit native woodland. Few of us are far from a hedge, although these too have been altered by new management practices and by invasion from non-native species. Fortunately there are still many miles of hedgerow made up of native trees, herbaceous plants and the animals that make these long, thin, woody places their homes.

One can learn much about the habits and names of many common trees, birds, insects and mammals by walking along a hedge; better still if the walk is done in every season of the year. By educating people to value and understand the habitats

9:14 The green-flowered alexanders (left) is not a native plant, unlike the white-flowered cow parsley (right)

9:15 Willow is grown extensively for biomass production

almost on their doorsteps, we may be investing intellectually in safeguards for the remote and inaccessible places and the plants and animals that make them special (as for example in the Burren of County Clare) – those places with a combination of species that makes them uniquely Irish.

In a century from now further changes to the Irish landscape will have been worked by natural forces and human hands. Together these will have shaped or re-shaped some part of the landscape, in the time-honoured manner. The story of Ireland's landscape has been, and continues to be, influenced by relentless change. At the beginning of the third millennium AD we cannot predict accurately the future for cereals, sugar beet, maize, silage, biomass fuel production or the biodiversity potential of set-aside land, still less the influence of urban expansion and the needs of the people of this island on its landscape and resources.

It is hoped that this text will have further informed people who love Ireland's wild places, the country's archaeology and its plants and animals, native and otherwise. It is nature and culture together that have shaped the landscape of this country and they have been inseparable for thousands of years. We all have a responsibility, individually and collectively, to understand how our Irish landscape developed and how we can care for it. These were the issues, always at the forefront of my mind, that prompted me to write this book.

BIBLIOGRAPHY

Aalen, F. H. A., Whelan, K. and Stout, M. (eds), *Atlas of the Irish Rural Landscape* (Cork, Cork University Press and University of Toronto Press, 1997)

— Whelan, K. and Stout, M. (eds), *Atlas of Ireland* (Dublin, Royal Irish Academy, 1979)

Ahlberg, B., Almgren, E., Wright, H. E., Ito, E. and Hobbie, S., 'Oxygen-isotope record of late-glacial climatic change in western Ireland', *Boreas* **25**, 257–67 (1996)

Ahlberg, K., Almgren, E., Wright, H. E. and Ito, E., 'Holocene stable-isotope stratigraphy at Lough Gur, County Limerick, Western Ireland', *The Holocene* **11**, 367–72 (2001)

Alexander, R., Coxon, P. and Thorn, R., 'A bog flow at Straduff Townland, County Sligo', *Proceedings of the Royal Irish Academy* **86B**, 107–119 (1986)

Andersen, S. Th., 'The differential pollen productivity of trees and its significance for the interpretation of a pollen diagram from a forested region', in H. J. B Birks and R. G. West, (eds), *Quaternary Plant Ecology* (Oxford, Blackwell, 1973)

— 'Identification of wild grass and cereal pollen', *Dansmark Geologiske Undersogelse Aarbog* **1978**, 69–92 (1979)

Andrieu, V., Huang, C. C., O'Connell, M. and Paus, A., 'Late-glacial vegetation and environment in Ireland: first results from four western sites', *Quaternary Science Reviews* **12**, 681–705 (1993)

Andrivon D., 'The origin of *Phytophthora infestans* populations present in Europe in the late 1840s: a critical review of historical and scientific evidence', *Plant Pathology* **45**, 1027–35 (1996)

ApSimon, A., 'The earlier Bronze age in the north of Ireland', *Ulster Journal of Archaeology* **32**, 28–72 (1969)

Baillie, M. G. L., 'Dendrochronology raises questions about the nature of the AD 536 dust-veil event', *The Holocene* **4**, 212–17 (1994)

— and Munro, M. A. R., 'Irish tree rings, Santorini and volcanic dust veils', *Nature* **332**, 344–6 (1988)

— and Brown, D. M., 'Some deductions on ancient Irish trees from dendrochronology', in J. R. Pilcher and S. S. Mac an tSaoir (eds), *Woods, Trees and Forests in Ireland*, 35–50 (Dublin, Royal Irish Academy, 1995)

Baker, A., Bolton, L., Brunsdon, C., Charlton, M. E. and McDermott, F., 'Visualisation of luminescence excitation-emission time-series: palaeoclimate implications from a 10,000 year stalagmite record from Ireland', *Geophysical Research Letters* **27**, 2145–8 (2000)

Baldini, J. U., McDermott, F. and Fairchild, I. J., 'Structure of the 8,200-year cold event revealed by speleothem trace-element record', *Science* **296**, 2203–6 (2002)

Barber, K. E., Chambers, F. M. and Maddy, D., 'Holocene palaeoclimates from peat stratigraphy: macrofossil proxy climate records from three oceanic raised bogs in England and Ireland', *Quaternary Science Reviews* **22**, 521–39 (2003)

Barnosky, A. D., 'Taphonomy and herd structure of the extinct Irish Elk, *Megaloceros giganteus*', *Science* **228**, 340–4 (1985)

— '"Big game" extinction caused by late Pleistocene climatic change: Irish elk (*Megaloceros giganteus*) in Ireland', *Quaternary Research* **25**, 128–35 (1986)

Barnosky, C. W., 'A late-glacial and post-glacial pollen record from the Dingle Peninsula, County Kerry', *Proceedings of the Royal Irish Academy* **88B**, 23–37 (1988)

Barry, T., *A History of Settlement in Ireland* (London, Routledge, 2000)

Battarbee, R. W., 'Observations on the recent history of Lough Neagh and its drainage basin', *Philosophical Transactions of the Royal Society of London* **B 281**, 303–45 (1978)

— 'Relative composition, concentration and calculated influx of diatoms from a sediment core from Lough Erne, Northern Ireland', *Polskie Archiwum Hydrobiologii* 25, 9–16 (1978)

Bayliss, A. and Woodman, P., 'A new Bayesian chronology for Mesolithic occupation at Mount Sandel, Northern Ireland', *Proceedings of the Prehistoric Society* **75**, 101–23 (2009)

Beirne, B. P., 'The distribution and origin of the British Lepidoptera', *Proceedings of the Royal Irish Academy* **49B**, 27–59 (1943)

Bell, J., 'A contribution to the study of cultivation ridges in Ireland', *Journal of the Royal Society of Antiquaries of Ireland* **114**, 80–97 (1984)

— and Watson, M., *Irish Farming: Implements and Techniques 1750–1900* (Edinburgh, John Donald, 1986)

Beltman, B. and Rouwenhorst, G., 'Chemical composition of rain in western Ireland', *Irish Naturalists' Journal* **24**, 267–74 (1993)

Bennett, K. D., 'Post-glacial history of *Pinus sylvestris* in the British Isles', *Quaternary Science Reviews* **3**, 133–55 (1984)
— 'Insularity and the Quaternary tree and shrub flora of the British Isles', in R. C. Preece (ed.), *Island Britain: a Quaternary Perspective*, 173–80 (London, The Geological Society, 1995)
— and Birks, H. J. B., 'Postglacial history of alder (*Alnus glutinosa* (L.) Gaertn.) in the British Isles', *Journal of Quaternary Science* **5**, 123–33 (1990)

Bennike, O., 'Colonisation of Greenland by plants and animals after the last ice age: a review', *Polar Record* **35**, 323–36 (1999)

Berglund, B., 'Human impact and climate changes – synchronous events and a causal link?', *Quaternary International* **105**, 7–12 (2003)

Biddles, R. D., *Use of Woodlands in the Late Anglo-Saxon Period* (London, Biddles, 1998)

Birks, H. J. B., 'Long-term ecological change in the British uplands', in M. B. Usher and D. B. A. Thompson (eds), *Ecological Change in the Uplands*, British Ecological Society Special Publication **7** (Oxford, Blackwell Scientific Publications, 1988)
— 'Holocene isochrone maps and the patterns of tree-spreading in the British Isles', *Journal of Biogeography* **16**, 503–40 (1989)
— Deacon, J. and Peglar, S., 'Pollen maps for the British Isles 5,000 years ago', *Proceedings of the Royal Society of London* **189**, 87–105 (1975)

Bjork, S., Rundgren, M., Ingolfsson, O. and Funder, S., 'The Preboreal oscillation round the Nordic Sea: terrestrial and lacustrine responses', *Journal of Quaternary Science* **12**, 455–65 (1987)

Bjorkman, L., 'The establishment of *Fagus sylvatica* at the stand-scale in southern Sweden', *The Holocene* **9**, 237–45 (1999)

Blackford, J. J., Edwards, K. E., Dugmore, A. J., Cook, G. T. and Buckland, P. C., 'Icelandic volcanic ash and the mid-Holocene Scots pine (*Pinus sylvestris*) decline in northern Scotland', *The Holocene* **2**, 260–5 (1992)

Blackford, J. J. and Chambers, F. M., 'Proxy climate record for the last 1,000 years from Irish blanket peat and a possible link to solar activity', *Earth and Planetary Science Letters* **133**, 145–50 (1995)

Bourke, A., *The Visitation of God? The Potato and the Great Irish Famine* (Dublin, Lilliput Press, 1993)

Bowen, D. Q., Philips, F. M., McCabe, A. M., Knutz, P. C. and Sykes, G. A., 'New data for the last Glacial Maximum in Great Britain and Ireland', *Quaternary Science Reviews* **21**, 89–101 (2002)

Bowler, M. and Bradshaw, R. H. W., 'Recent accumulation and erosion of blanket peat in the Wicklow Mountains, Ireland', *New Phytologist* **101**, 543–50 (1985)

Boyle, E. A., 'Is ocean thermohaline circulation linked to abrupt stadial/interstadial transitions?', *Quaternary Science Reviews* **19**, 255–72 (2000)

Bradshaw, R. H. W., 'Palaeoecology: a general review', in R. H. Thorn (ed.), *Sligo and West Leitrim*, Field Guide 8 (Dublin, Irish Association for Quaternary Studies, 1985)
— and Browne, P., 'Changing patterns in the post-glacial distribution of *Pinus sylvestris* in Ireland', *Journal of Biogeography* **14**, 237–48 (1987)
— and McGee, E., 'The extent and time-course of mountain peat erosion in Ireland', *New Phytologist* **108**, 219–24 (1988)
— and Hannon, G. E., 'Human influence in the decline of the Irish forest ecosystem', in F. Salbitano (ed.), *Human Influence on Forest Ecosystems Development in Europe* (Bologna, Pitagora Editrice, 1988)
— and Mitchell, F. J. G., 'The palaeoecological approach to reconstructing former grazing-vegetation interactions', *Forest Ecology and Management* **120**, 3–12 (1999)
— and Hannon, G. E., 'The Holocene structure of north-west European temperate forest induced from palaeoecological data', in O. Honnay, K. Verheyen, K. Bossuyt, and M. Hermy (eds), *Forest Biodiversity: Lessons from History for Conservation*, 11–23 (Oxford, CAB International, 2001)

Bramwell, D., 'The birds of Britain: when did they arrive?', in R. D. S. Jenkinson and D. D. Gilbertson (eds), *In the Shadow of Extinction*, 90–4 (Sheffield, Collos, 1984)

Broecker, W. S. and Denton, G. H., 'What drives glacial cycles?', *Scientific American* **262**, 42–50 (1990)

Brothwell, D., 'Variation in early Irish populations: a brief survey of the evidence', *Ulster Journal of Archaeology* **48**, 5–9 (1985)

Brown, A. G., 'Clearances and clearings: deforestation in Mesolithic/Neolithic Britain', *Oxford Journal of Archaeology* **16**, 133–46 (1997)
— 'Biodiversity and pollen analysis: modern pollen studies and the recent history of a flood plain in SW Ireland', *Journal of Biogeography* **26**, 19–32 (1999)
— Hatton, J., O'Brien, C., Selby, K., Langdon, P., Stuijts, I. and Caseldine, C., 'Vegetation, landscape and human activity in Midland Ireland; mire and lake records from Lough Kinale–Derragh Lough area, central Ireland', *Vegetation History and Archaeobotany* **14**, 81–9 (2005)

Bryant, R. H., 'A late-Midlandian section at Finglas river, near Waterville, Kerry', *Proceedings of the Royal Irish Academy* **74B**, 161–78 (1974)

Bush, M. B. and Hall, A. R., 'Flandrian *Alnus*: expansion or immigration?', *Journal of Biogeography* **14**, 479–81 (1987)

Butler, C. J., Coughlin, A. D. S. and Fee, D. T., 'Precipitation at Armagh Observatory 1838–1997', *Proceedings of the Royal Irish Academy* **98B**, 123–40 (1998)
— Suarez, A. M. G., Coughlin, A. D. S. and Morrell, C., 'Air temperatures at Armagh Observatory, Northern Ireland, from 1796 to 2002', *International Journal of Climatology* **25**, 1055–79 (2005)

Buzer, J. S., 'Pollen analyses of sediments from Lough Ine and Ballyally Lough, Co. Cork, SW Ireland', *New Phytologist* **86**, 93–108 (1975)
— 'Diatom analyses of sediments from Lough Ine, Co. Cork, SW Ireland', *New Phytologist* **89**, 511–33 (1981)

Calendar of Carew Manuscripts 1603–1623, 150 (London, 1873)

Calendar of State Papers 1600–1601, 92–3 (London, 1905)

Carlisle, A. and Brown, A. H. F., '*Pinus sylvestris* L.', *Journal of Ecology* **56**, 269–307 (1968)

Carter, R. W. G., 'Recent variations in sea-level on the north and east coasts of Ireland and associated shoreline response', *Proceedings of the Royal Irish Academy* **82B**, 177–87 (1982)
— 'Raised coastal landforms as products of modern process variations, and their relevance in eustatic sea-level studies: examples from eastern Ireland', *Boreas* **12**, 167–82 (1983)
— Devoy, R. J. N. and Shaw, J., 'Late Holocene sea levels in Ireland', *Journal of Quaternary Science* **4**, 7–24 (1989)

Case, H., 'Settlement-patterns in the north Irish Neolithic', *Ulster Journal of Archaeology* **32**, 3–27 (1969)

Caseldine, C. and Hatton, J., 'Early land clearance and wooden trackway construction in the third and fourth millennia BC at Corlea, Co. Longford', *Proceedings of the Royal Irish Academy, Biology and Environment* **96B**, 11–19 (1996)
— Hatton, J., Huber, U., Chiverrell, R. and Woolley, N., 'Assessing the impact of volcanic activity on mid-Holocene climate on Ireland; the need for replicate data', *The Holocene* **8**, 105–11 (1998)
— and Gearey, B., 'A multiproxy approach to reconstructing surface wetness changes and prehistoric bog bursts in a raised mire at Derryville, Co. Tipperary, Ireland', *The Holocene* **15**, 585–601 (2005)
— Thompson, G., Langdon, C. and Hendon, D., 'Evidence for an extreme climatic event on Achill Island, Co. Mayo, Ireland around 5200–5100 cal. yr BP', *Journal of Quaternary Science* **20**, 169–78 (2005)

Caulfield, S., 'Neolithic fields: the Irish evidence', in H. C. Bowen and P. J. Fowler (eds), *Early Land Allotment in the British Isles. A Survey of Recent Work*, British Archaeological Reports, British Series **48** (Oxford, 1978)
— 'The Neolithic settlement of north Connaught', in T. Reeves-Smyth and F. Hamond (eds), *Landscape Archaeology in Ireland*, British Archaeological Reports, British Series **116** (Oxford, 1983)
— O'Donnell, R. G. and Mitchell, P. I., 'Radiocarbon dating of a Neolithic field system at Céide Fields, Co. Mayo, Ireland', *Radiocarbon* **40**, 629–40 (1998)

Chambers, F. M. and Elliot, L., 'Spread and expansion of *Alnus* Mill. in the British Isles: timing, agencies and possible vectors', *Journal of Biogeography* **16**, 541–50 (1989)

Charles-Edwards, T. and Kelly, F. (eds), *Bechbretha: an Old Irish Law-tract on Bee-keeping* (Dublin, Institute of Advanced Studies, 1983)

Charlesworth, J. K., 'Some geological observations on the origin of the Irish fauna and flora', *Proceedings of the Royal Irish Academy* **39B**, 358–90 (1930)

Clark, P. U., Marshall, S. J., Clarke, G. K. C., Hostetler, S. W., Licciardi, J. M. and Teller, J. T., 'Freshwater forcing of abrupt climate change during the last glaciation', *Science* **293**, 283–7 (2001)

Clarkson, L. A. and Crawford, E. M., *Feast and Famine. Food and Nutrition in Ireland 1500–1920* (Oxford, Oxford University Press, 2001)

Coard, R. and Chamberlain, A. T., 'The nature and timing of faunal change in the British Isles across the Pleistocene/Holocene transition', *The Holocene* **9**, 372–6 (1999)

Cole, E. E. and Mitchell, F. J. G., 'Human impact on the Irish landscape during the late Holocene inferred from palynological studies at three peatland sites', *The Holocene* **13**, 507–15 (2003)

Coles, J. M., Heal, S. V. E. and Orme, B.J., 'The use and character of wood in prehistoric Britain and Ireland', *Proceedings of the Prehistoric Society* **44**, 1–45 (1978)

Colhoun, E. A., Dickson, J. H., McCabe, A. M. and Shotton, F. W., 'A middle Midlandian freshwater series at Derryvree, Maguiresbridge, County Fermanagh, Ireland', *Proceedings of the Royal Society of London* **180**, 273–92 (1972)
— and McCabe A. M., 'Pleistocene glacial, glaciomarine and associated deposits of Mell and Tullyallen townlands, near Drogheda, eastern Ireland', *Proceedings of the Royal Irish Academy* **73B**, 165–207 (1973)
— and Synge, F. M., 'The cirque moraines at Lough Nahanagan, County Wicklow, Ireland', *Proceedings of the Royal Irish Academy* **80B**, 25–45 (1980)

Collins, A. E. P., 'Excavations at Lough Faughan crannóg, Co. Down 1951–1952', *Ulster Journal of Archaeology* **18**, 45–82 (1955)

Collins, T. and Coyne, F., 'Fire and water; early Mesolithic cremations at Castleconnell, Co. Limerick', *Archaeology Ireland* **17**, 24–7 (2003)

Cooney, G., *Landscapes of Neolithic Ireland* (London, Routledge, 2000)

Coope, G. R., 'Fossil coleopteran assemblages as sensitive indicators of climatic changes during the Devensian (Last) cold stage', *Philosophical Transactions of the Royal Society of London* **B 280**, 313–40 (1977)
— Dickson, J. H., McCutcheon, J. A. and Mitchell, G. F., 'The lateglacial and early postglacial deposit at Drumurcher, Co. Monaghan', *Proceedings of the Royal Irish Academy* **79B**, 63–85 (1979)
— Lemdhl, G., Lowe, J. J. and Walking, A., 'Temperature gradients in northern Europe during the last glacial-Holocene transition (14–9 ^{14}C kyr bp) interpreted from coleopteran assemblages', *Journal of Quaternary Science* **13**, 419–34 (1998)

Cooper, J. A. G., Kelley, J. T., Belknap, D. F., Quinn, R. and McKenna, J., 'Inner shelf seismic stratigraphy off the north coast of Northern Ireland: new data on the depth of the Holocene lowstand', *Marine Geology* **186**, 369–87 (2002)

Corbet, G. B., 'Origin of the British insular races of small mammals and of the "Lusitanian" fauna', *Nature* **191**, 1037–40 (1961)

Coxon, P., *Clare Island, Co. Mayo*, Field Guide **5** (Dublin, Irish Association for Quaternary Studies, 1982)
— 'A radiocarbon dated early post-glacial pollen diagram from a pingo remnant near Millstreet, Co. Cork', *Irish Journal of Earth Sciences* **8**, 9–20 (1986)
— 'A post-glacial pollen diagram from Clare Island, Co. Mayo', *Irish Naturalists' Journal* **22**, 217–23 (1987)
— and Waldren, S., 'The floristic record of Ireland's Pleistocene temperate stages', in R. C. Preece (ed.), *Island Britain: a Quaternary Perspective*, 243–67 (London, The Geological Society, 1995)
— and Waldren, S., 'Flora and vegetation of the Quaternary temperate stages of NW Europe', in B. Huntley, W. Cramer, A. V. Morgan, I. C. Prentice and J. R. M. Allen (eds), *Evidence from Large-scale Range Changes. Past and Future Rapid Environmental Changes: the Spatial and Evolutionary Responses of Terrestrial Biota*, NATO ASI Series 1, Global Environmental Change 47 (Berlin, Springer Verlag, 1997)

Craig, A. J., 'Pollen percentage and influx analysis in south-west Ireland: a contribution to the ecological history of the Late-glacial period', *Journal of Ecology* **66**, 297–324 (1978)

Culleton, E., *Celtic and Early Christian Wexford, AD 400–1166* (Dublin, Four Courts Press, 1999)
— and Mitchell, G. F., 'Soil erosion following deforestation in the early Christian period of South Wexford', *Journal of the Royal Society of Antiquaries of Ireland* **106**, 120–3 (1976)

Cunliffe, B., *Facing the Ocean: the Atlantic and its Peoples* (Oxford, Oxford University Press, 2004)

Currant, A. and Jacobi, R., 'Vertebrate fauna of the British Late Pleistocene and the chronology of human settlement', *Quaternary Newsletter* **82**, 1–8 (1997)

Curtis, T. G. F. and Sheehy Skeffington, M. J., 'The salt marshes of Ireland: an inventory and account of their geographical variation', *Proceedings of the Royal Irish Academy, Biology and Environment* **98B**, 87–104 (1998)

Cwynar, L. C. and Watts, W. A., 'Accelerator-mass spectrometer ages for late-glacial events at Ballybetagh, Ireland', *Quaternary Research* **31**, 377–80 (1989)

D'Arcy, G., 'Little bird bone: long story', *Irish Wildlife*, summer edition, 10–12 (2006)

Davenport, J. L., Sleeman, D. P. and Woodman, P. C. (eds), 'Mind the Gap: postglacial colonization of Ireland', *Irish Naturalists' Journal*, special supplement (2008)

Deacon, J., 'The location of refugia of *Corylus avellana* L. during the Weichselian glaciation', *New Phytologist* **73**, 1055–63 (1974)

De Groot, W. J, Thomas, P. A. and Wein, R. W., '*Betula nana* L., *Betula glandulosa* Michx', *Journal of Ecology* **85**, 241–64 (1997)

Devoy, R. J., 'The problem of a late Quaternary landbridge between Britain and Ireland,' *Quaternary Science Reviews* **4**, 43–58 (1985)

— 'Deglaciation, Earth crustal behaviour and sea-level changes in the determination of insularity: a perspective from Ireland', in R. C. Preece (ed.), *Island Britain: a Quaternary Perspective*, 109–120 (London, The Geological Society, 1995)

Dickson, D., *Arctic Ireland: the Extraordinary Story of the Great Frost and Forgotten Famine of 1740–41* (Belfast, White Row Press, 1998)

Dodson, J. R., 'Fine resolution pollen analysis of vegetation history in the Lough Adoon Valley, Co. Kerry, western Ireland', *Review of Palaeobotany and Palynology* **64**, 235–45 (1990)

— 'The Holocene vegetation of a prehistorically inhabited valley, Dingle Peninsula, Co. Kerry', *Proceedings of the Royal Irish Academy* **90B**, 151–74 (1990)

— and Bradshaw, R. H. W., 'A history of vegetation and fire, 6000 BP to present', *Boreas* **16**, 113–23 (1987)

Doherty, C., 'Exchange and trade in early medieval Ireland', *Journal of the Royal Society of Antiquaries of Ireland* **110**, 67–89 (1980)

Dowling, L. A. and Coxon, P., 'Current understanding of Pleistocene temperate stages in Ireland', *Quaternary Science Reviews* **20**, 1631–42 (2001)

Downey, C. and Singh, G., 'Dinoflagellate cysts from estuarine and raised beach deposits at Woodgrange, Co. Down, N. Ireland', *Grana Palynologica* **9**, 1–3 (1969)

Downey, L., Synnott, C., Kelly, E. P. and Stanton, C., 'Bog butter: dating profile and location', *Archaeology Ireland* **20**, 32–4 (2006)

Dresser, P. Q., Smith, A. G. and Pearson, G. W., 'Radiocarbon dating of the raised beach at Woodgrange, Co. Down', *Proceedings of the Royal Irish Academy* **73B**, 53–6 (1973)

Dugmore, A. J., 'Icelandic volcanic ash in Scotland', *Scottish Geographical Magazine* **1055**, 168–72 (1989)

Dwyer, R. B and Mitchell, F. J. G., 'Investigation of the environmental impact of remote volcanic activity on north Mayo, Ireland, during the mid-Holocene', *The Holocene* **7**, 113–18 (1997)

Edwards C., 'Extinct giant deer traced to modern relative by ancient DNA', *IQUA Newsletter* **35**, 3 (2005)

Edwards, K. J., 'Vegetational changes associated with man in Co. Tyrone', in K. J. Edwards (ed.), *Co. Tyrone, Field Guide* **3** (Dublin, Irish Association for Quaternary Studies, 1980)

— 'Man, space and the woodland edge: speculations on the detection and interpretation of human impact on pollen profiles', in M. Bell and S. Limbrey (eds), *Archaeological Aspects of Woodland Ecology*, British Archaeological Reports, International Series **146** (Oxford, 1982)

— and Hirons, K. R., 'Cereal pollen grains in pre-elm decline deposits: implications for the earliest agriculture in Britain and Ireland', *Journal of Archaeological Science* **11**, 71–80 (1984)

— and Warren, W. P. (eds), *The Quaternary History of Ireland* (London, Academic Press, 1985)

— and Whittington, G., 'Lake sediments, erosion and landscape change during the Holocene in Britain and Ireland', *Catena* **42**, 143–73 (2001)

Erdtman, G., 'Trace of the history of the forests of Ireland', *Irish Naturalists' Journal* **1**, 242–25 (1927)

Faegri, K. and Iversen, J., *Textbook of Pollen Analysis*, edn 4 (Chichester, Wiley & Son, 1989)

Fairley, J. S., *An Irish Beast Book* (Belfast, The Blackstaff Press, 1975)

Farrington, A., 'Organisation of the committee for Quaternary research in Ireland', *Irish Naturalists' Journal* **5**, 128–30 (1934)

Feehan, J., *Farming in Ireland. History, Heritage and Environment* (Dublin, University College, 2003)
— and O'Donovan, G., *The Bogs of Ireland; an Introduction to the Natural, Cultural and Industrial Heritage of Irish Peatlands* (Dublin, The Environment Institute, University College, 1996)

Fisher, N., 'The last Irish wolf', *Irish Naturalists' Journal* **5**, 41 (1934)

Forbes, A. C., 'Some legendary and historical references to Irish woods, and their significance', *Proceedings of the Royal Irish Academy* **41B**, 15–36 (1932)

Foss, P. J. and Doyle, G. J., 'The history of *Erica erigena* (R. Ross), an Irish plant with a disjunct European distribution', *Journal of Quaternary Science* **5**, 1–16 (1990)
— Doyle, G. J. and Nelson, E. C., 'The distribution of *Erica erigena* R. Ross in Ireland', *Watsonia* **16**, 311–27 (1987)

Fossitt, J. A., 'Late-glacial and Holocene vegetation history of western Donegal, Ireland', *Proceedings of the Royal Irish Academy, Biology and Environment* **94B**, 1–31 (1994)

Freeman, T. W., *Ireland: a General and Regional Geography*, edn 4 (London, Methuen, 1972)

Frendengren, C., McClatchie, M. and Stuijts, I., 'Connections and distance, investigating social and agricultural issues relating to early medieval crannógs in Ireland', *Environmental Archaeology* **9**, 173–8 (2004)

Garbett, G. G., 'The elm decline: the depletion of a resource', *New Phytologist* **88**, 573–85 (1981)

Gear, A. and Huntley, B., 'Rapid changes in the range limits of Scots pine 4,000 years ago', *Science* **251**, 544–7 (1991)

Geel, B. van and Middeldorp, A., 'Vegetational history of Carbury bog (Co. Kildare, Ireland) during the last 850 years and a test of temperature indicator value of 2H/1H measurements of peat samples in relation to the historical sources and meteorological data', *New Phytologist* **109**, 377–92 (1988)

Gennard, D. E. and Hackney, C. R., 'First Irish record of a fossil bracket fungus *Fomes fomes* (L. EX FR.) KICKX', *Irish Naturalists' Journal* **23**, 19–21 (1989)

Gibson, C. E., Anderson, N. J., Zhou, Q., Allen, M. and Appleby, P. G., 'Changes in sediment and diatom deposition in Lower Lough Erne *c.* 1920–90', *Proceedings of the Royal Irish Academy, Biology and Environment* **103B**, 31–39 (2003)

Girling, M. A. and Greig, J. R. A., 'A first record for *Scolytus scolytus* (F.) (elm bark beetle): its occurrence in elm decline deposits from London and the implications for Neolithic elm decline', *Journal of Archaeological Science* 12, 347–51 (1985)

Godwin, H., *The History of the British Flora*, edn 2 (Cambridge, Cambridge University Press, 1975)

Goransson, H., 'Pollen analytical investigations in Cloverhill Lough, Carrowmore, Co. Sligo, Ireland', in B. Burenhult (ed.), *The Carrowmore Excavations, Excavation Season 1980*, Stockholm Archaeological Reports **7**, 125–39 (Stockholm, Institute of Archeology, University of Stockholm, 1980)
— 'Pollen analytical investigations in Ballygawley Lough and Carrowkeel, Co. Sligo, Ireland', in B. Burenhult (ed.), *The Carrowmore Excavations, Excavation Season 1981*, Stockholm Archaeological Reports **8**, 180–95 (Stockholm, Institute of Archeology, University of Stockholm, 1981)
— 'Pollen analytical investigations in the Sligo area', in G. Burenhult (ed.), *The Archaeology of Carrowmore*, 154–93 (Stockholm, Institute of Archaeology, University of Stockholm, 1984)

Gosler, T., Arnold, M. and Padur, M. F., 'The Younger Dry as cold event – was it synchronous over the North Atlantic region?', *Radiocarbon* **37**, 63–70 (1995)

Graham, J. M., 'South-west Donegal in the seventeenth century', *Irish Geography* **6**, 136–52 (1970)

Groenman-van-Waatering, W., 'The origin of crop weed communities composed of summer annuals', *Vegetatio* **41**, 57–9 (1979)

— 'The early agricultural utilization of the landscape: the last word on the elm decline?', in T. Reeves-Smith and F. Hamond (eds), *Landscape Archaeology in Ireland*, British Archaeological Reports, British Series **116**, 217–32 (Oxford, 1983)

— and Pals, J. P., 'Pollen and seed analysis', in M. J. O'Kelly (ed.), *Newgrange, Archaeology, Art and Legend*, 219–23 (London, Thames & Hudson, 1982)

Guthrie, R. D., 'Palaeolithic art as resource in artiodactyl palaeobiology', in E.S. Vrba and G. B. Schaller (eds), *Antelopes, Deers and Relatives – Fossil Record, Behavioural Ecology, Systematics and Conservation*, 96–127 (London, Yale University Press, 2000)

Hackney, P. (ed.), *Stewart and Corry's Flora of the north-east of Ireland*, edn 3 (Belfast, Institute of Irish Studies, Queen's University, 1992)

Hall, V. A., 'The historical and palynological evidence for flax cultivation in mid Co. Down', *Ulster Journal of Archaeology* **52**, 5–10 (1989)

— 'A study of the modern pollen rain from a reconstructed nineteenth century farm', *Irish Naturalists' Journal* **23**, 82–92 (1989)

— 'The vegetational landscape of mid Co. Down over the last half millennium', *Ulster Folklife* **35**, 72–85 (1989)

— 'Ancient agricultural activity at Slieve Gullion, Co. Armagh: the palynological and documentary evidence', *Proceedings of the Royal Irish Academy* **90C**, 123–34 (1990)

— 'Recent landscape history from a Co. Down lake deposit', *New Phytologist* **115**, 377–83 (1990)

— 'Detecting re-deposited pollen in an Irish lake deposit', *Irish Naturalists' Journal* **23**, 398–402 (1991)

— 'The woodlands of the Lower Bann valley in the seventeenth century; the documentary evidence', *Ulster Folklife* **38**, 1–11 (1992)

— 'Landscape development in NE Ireland over the last three centuries', *Review of Palaeobotany and Palynology* **82**, 75–82 (1994)

— 'Woodland depletion in Ireland over the last millennium', in J.R. Pilcher and S.S. Mac an tSaoir (eds), *Woods, Trees and Forests in Ireland*, 23–34 (Dublin, Royal Irish Academy, 1995)

— 'Recent landscape change and landscape reconstruction in Northern Ireland: a tephra-dated pollen study', *Review of Palaeobotany and Palynology* **103**, 59–68 (1998)

— 'Vegetation history of mid- to western-Ireland in the second millennium AD; fresh evidence from tephra-dated palynological investigations', *Vegetation History and Archaeobotany* **12**, 7–17 (2003)

— Pilcher, J. R. and Bowler, M., 'Pre-elm decline cereal-size pollen; evaluating its recruitment to fossil deposits using modern pollen rain studies', *Proceedings of the Royal Irish Academy, Biology and Environment* **1**, 1–4 (1993)

— Pilcher, J. R. and McCormac, F. G., 'Tephra dated lowland landscape history of the north of Ireland AD 750–1150', *New Phytologist* **125**, 193–202 (1993)

— Pilcher, J. R. and McCormac, F. G., 'Icelandic volcanic ash and the mid-Holocene pine (*Pinus sylvestris*) pollen decline in the north of Ireland; no correlation', *The Holocene* **4**, 79–83 (1994)

— and Pilcher, J. R., 'Irish grassland history', in D. W. Jeffrey, M. B. Jones and J. H. McAdam (eds), *Irish Grasslands – their Biology and Management*, 188–93 (Dublin, Royal Irish Academy, 1995)

— and Mauquoy, D., 'Tephra-dated climate and human impact studies during the last 1,500 years from a raised bog in central Ireland', *The Holocene* **15**, 1086–93 (2005)

Hannon, G. E. and Bradshaw, R. H. W., 'Recent vegetation dynamics on two Connemara lake islands, western Ireland', *Journal of Biogeography* **16**, 75–81 (1989)

Harrison, S. and Mighall, T. M. (eds), *The Quaternary of South West Ireland* (London, Quaternary Research Association, 2002)

Head, K., Turney, C. S. M., Pilcher, J. R., Palmer, J. G. and Baillie, M. G. L., 'Problems with identifying the "8200-year cold event" in terrestrial records of the Atlantic seaboard: a case study from Dooagh, Achill Island, Ireland', *Journal of Quaternary Science* **22**, 65–75 (2007)

Heery, S., *The Shannon Floodlands; a Natural History* (Kinvara, Tir Eolas, 1993)

Heybroek, H. M., 'Diseases and lopping of fodder as possible causes of prehistoric decline of *Ulmus*', *Acta Botanica Neerlandica* **12**, 1–11 (1963)

Hickie, D. and O'Toole, M., *Native Trees and Forests of Ireland* (Dublin, Gill & Macmillan, 2002)

Hill, G., *An Historical Account of the Plantation of Ulster at the Commencement of the Seventeenth Century 1608–1620* (Shannon, Irish University Press facsimile reprint 1970, first published 1877)

Hill, A. R. and Prior, D. B., 'Directions of ice movement in north-east Ireland', *Proceedings of the Royal Irish Academy* **66B**, 71–84 (1968)

Hirons, K. R., 'Percentage and accumulation rate pollen diagrams from east Co. Tyrone', in T. Reeves-Smith and F. Hamond (eds), *Landscape Archaeology in Ireland*, British Archaeological Reports, British Series **116**, 95–117 (Oxford, 1983)
— 'Recruitment of cpr2 pollen to lake sediments: an example from Co. Tyrone, Northern Ireland', *Review of Palaeobotany and Palynology* **54**, 43–54 (1988)
— and Edwards, K. J., 'Events at and around the first and second *Ulmus* declines; palaeoecological investigations in Co. Tyrone, Northern Ireland', *New Phytologist* **104**, 131–53 (1986)

Holm, P., 'The slave trade of Dublin, ninth to twelfth century', *Peritia* **5**, 317–45 (1996)

Hore, H., 'Marshall Begenal's description of Ulster; Anno 1586', *Ulster Journal of Archaeology* **3**, 137–45 (1856)
— 'Woods and fastnesses in ancient Ireland', *Ulster Journal of Archaeology* **3**, 145–61 (1856)

ten Hove,, H. A., 'The *Ulmus* fall at the transition Atlanticum-Subboreal in pollen diagrams', *Palaeogeography, Palaeoclimatology, Palaeoecology* **5**, 359–69 (1968)

Huang, C. C., 'Holocene landscape development and human impact in the Connemara uplands, western Ireland', *Journal of Biogeography* **29**, 153–65 (2002)
— and O'Connell, M., 'Recent land-use history in eastern Connemara, a palaeoecological case study at Ballydoo Lough, Connemara, Co. Galway', in J. Feehan (ed.), *Environment and Development in Ireland* (Dublin, Environment Institute, 1991)
— and O'Connell, M., 'Recent land-use and soil erosion history within a small catchment in Connemara, western Ireland: evidence from lake sediments and documentary evidence', *Catena* **41**, 293–335 (2000)

Hulme, P. E., 'Natural regeneration of yew (*Taxus baccata* L.): microsite, seed or herbivore limitation?', *Journal of Ecology* **84**, 853–61 (1996)

Huntley, B., 'European post-glacial forests: compositional changes in response to climatic change', *Journal of Vegetation Science* **1**, 507–18 (1990)
— 'Rapid early-Holocene migration and high abundance of hazel (*Corylus avellana* L.), alternative hypotheses', in F. M. Chambers (ed.), *Climate Change and Human Impact on the Landscape*, 205–16 (London, Chapman & Hall, 1993)
— and Birks, H. J. B., *An Atlas of Past and Present Pollen Maps for Europe 0–13,000 years ago* (Cambridge, Cambridge University Press, 1983)

Igoe, F. and Hammar, J., 'The Arctic Char (*Salvelinus alpinus* L.) species complex in Ireland: a secretive and threatened ice age relict', *Proceedings of the Royal Irish Academy, Biology and Environment* **3**, 73–92 (2004)

Innes, J. B. and Blackford, J. J., 'The ecology of late Mesolithic woodland disturbances: model testing with fungal spore assemblages', *Journal of Archaeological Science* **30**, 185–94 (2003)
— Chiverrell, R. C., Blackford, J. J., Davey, P. J., Gonzalez, S., Rutherford, M. M. and Tomlinson, P. R., 'Earliest Holocene vegetation history and island biogeography of the Isle of Man, British Isles', *Journal of Biogeography* **31**, 761–72 (2004)

Iversen, J., 'Late- and post-glacial forest history in central Europe', *Oikos* **2**, 315–18 (1950)
— 'Problems of the early post-glacial forest development in Denmark', *Danmarks Geologiske Undersogelse* **3**, 1–32 (1960)

Jacobson, G. L. Jr and Bradshaw, R. H. W., 'The selection of sites for palaeovegetational studies', *Quaternary Research* **16**, 80–96 (1981)

Jalas, J. and Suominen, J., *Atlas Flora Europaea*, 9 vols (incomplete) (Helsinki, Committee for Mapping the Flora of Europe, 1976–1991)

Jelicic, L. and O'Connell, M., 'History of vegetation and land use from 3200 B.P. to the present in the north-west Burren, a karstic region of western Ireland', *Vegetation History and Archaeobotany* **1**, 119–40 (1992)

Jessen, K., 'Preliminary report on bog investigations in Ireland, 1934', *Irish Naturalists' Journal* **5**, 130–4 (1934)

— 'Studies in late Quaternary deposits and flora-history of Ireland', *Proceedings of the Royal Irish Academy* **52B**, 85–290 (1949)

— and Farrington, A., 'The bogs at Ballybetagh, near Dublin, with remarks on Late-Glacial conditions in Ireland', *Proceedings of the Royal Irish Academy* **44B**, 205–60 (1938)

Johansen, S. and Hytteborn, H., 'A contribution to the discussion of biota and dispersal with drift ice and driftwood in the North Atlantic', *Journal of Biogeography* **28**, 105–15 (2001)

Johnson, M. P. and Simberloff, D. S., 'Environmental determinants of island species number in the British Isles', *Journal of Biogeography* **1**, 149–54 (1974)

Jones, G., Ll. and McKeever, M., 'The sedimentology and palynology of some postglacial deposits from Marble Arch Caves, Co. Fermanagh', *Cave Science* **14**, 3–6 (1987)

Juvigne, E., Bastin, B., Delibrias, G., Evin, J. and Streel, M., 'A method of estimating the migration time of plant species within the time range of ^{14}C-dating', *Quaternary International* **47/48**, 147–52 (1998)

Kelly, F., 'The old Irish tree-list', *Celtica* **11**, 107–24 (1976)

— *A Guide to Early Irish Law* (Dublin, Institute of Advanced Studies, 1988)

— *Early Irish Farming* (Dublin, Institute of Advanced Studies, 1997)

Kerney, M. P., 'Early Post-glacial deposits in King's County, Ireland, and their molluscan fauna', *Journal of Conchology* **24**, 155–64 (1957)

Kerr, T. R., *Early Christian Settlement in North-west Ulster*, British Archaeological Reports, British Series **430** (Oxford, 2007)

— Swindells, G. T. and Plunkett, G., 'Making hay while the sun shines? Socio-economic change, cereal production and climatic deterioration in Early Medieval Europe', *Journal of Archaeological Science* **36**, 2868–74 (2009)

Kertland, M. P. H., 'The cloudberry (*Rubus chamaemorus* L.) in County Tyrone', *Irish Naturalists' Journal* **12**, 309–14 (1958)

Kinahan, G. H., *Manual of the Geology of Ireland* (London, Keegan Paul & Co., 1878)

King, A. K. L. and Morrison, M. E. S., '*Sphagnum imbricatum*, HORNSCH. EX.RUSS', *Irish Naturalists' Journal* **12**, 105–7 (1956)

Kirk, S. M., 'High altitude cereal growing in County Down, Northern Ireland? A note', *Ulster Journal of Archaeology* **36**, 99–100 (1973)

Krog, H., 'Pollen analytical investigation of a ^{14}C-dated Allerod section from Ruds Vedby', *Danmarks Geologiske Undersogelse* **11**, 120–39 (1954)

Kukla, G. and Gavin, J., 'Did glacials start with global warming?', *Quaternary Science Reviews* **24**, 1547–57 (2005)

Kutzbach, J. E. and Ruddiman, W. F., 'Model description, external forcing, and surface boundary conditions', in H. E. Wright Jr, J. E. Kutzbach, T. Webb III, W. F. Ruddiman, F. A. Street-Perrott and P. J. Bartlein (eds), *Global Climates since the Last Glacial Maximum*, 12–23 (Minneapolis, University of Minnesota Press, 1993)

Lambeck, K., 'Glaciation and sea-level change for Ireland and the Irish Sea since Late Devensian/Midlandian time', *Journal of the Geological Society of London* **153**, 853–72 (1996)

— and Purcell, A. P., 'Sea-level change in the Irish Sea since the Last Glacial Maximum: constraints from isostatic modelling', *Journal of Quaternary Science* **16**, 497–506 (2001)

Larsen, C. S., 'The agricultural revolution as environmental catastrophe: implications for health and lifestyle in the Holocene', *Quaternary International* **150**, 12–20 (2006)

Lister, A. M., 'The evolution of the giant deer, *Megaloceros giganteus* (Blumenbach)', *Zoological Journal of the Linnean Society* **112**, 65–100 (1994)

Little, D. J., Mitchell, F. J. G., von Engelbrechten, S. and Farrell, E. P., 'Assessment of impact of past disturbance and prehistoric *Pinus sylvestris* on vegetation dynamics and soil development in Uragh Wood, SW Ireland', *The Holocene* **6**, 90–9 (1996)

Lomas-Clarke, S. and Barber, K., 'Palaeoecology of human impact during the historic period: palynology and geochemistry of a peat deposit at Abbeyknockmoy, Co. Galway, Ireland', *The Holocene* **14**, 721–31 (2004)

Lowe, J. J., Birks, H. H., Brooks, S. J., Coope, G. R., Harkness, D. D., Mayle, F. E., Sheldrick, C., Turney, C. S. M. and Walker, M. J. C., 'The chronology of palaeoenvironmental changes during the last Glacial-Holocene transition: towards an event stratigraphy for the British Isles', *Journal of the Geological Society, London* **156**, 397–410 (1999)

Lucas, A. T., 'Prehistoric block-wheels from Doogarymore, Co. Roscommon and Timahoe East, Co. Kildare', *Journal of the Royal Society of Antiquaries of Ireland* **102**, 19–48 (1972)
— *Cattle in Ancient Ireland* (Kilkenny, Boethius, 1989)

Lynch, A., *Man and Environment in South-west Ireland, 4000 BC–AD 800. A Study of Man's Impact on the Development of Soil and Vegetation*, British Archaeological Reports, British Series **85** (1981)

Lynn, C. J., 'Trial excavations at the King's Stables, Tray townland, County Armagh. Appendix 1, Penn, C. An osteological analysis of the animal remains from the King's Stables', *Ulster Journal of Archaeology* **40**, 42–62 (1977)

Mahr, A., 'Quaternary research in Ireland. 1934, from an archaeological viewpoint', *Irish Naturalists Journal* **5**, 137–44 (1934)

Mallory, J. P. and Baillie, M. G. L., 'The fall of the house of oak', *Emania* **6**, 27–33 (1988)
— and McCormick, F., 'Excavations at Ballymulholland, Magilligan Foreland, Co. Londonderry', *Ulster Journal of Archaeology* **51**, 103–14 (1988)

Maloney, B. K., 'A palaeoecological investigation of the Holocene back-beach barrier environment near Carnsore Point, Co. Wexford', *Proceedings of the Royal Irish Academy* **85B**, 73–89 (1985)

Manning, C., 'The 1653 survey of the lands granted to the countess of Ormond in Co. Kilkenny', *Journal of the Royal Society of Antiquaries of Ireland* **129**, 40–66 (1999)

Martinova, N., McDonald, R. A. and Searle, J. B., 'Stoats (*Mustela erminea*) provide evidence of natural overland colonization of Ireland', *Proceedings of the Royal Society* **B 274**, 1387–93 (2007)

Matthews, J. R., *Origin and Distribution of the British Flora* (London, Hutchinson, 1995)

McAuley, I. R. and Watts, W.A., 'Dublin radiocarbon dates', *Radiocarbon* **3**, 26–8 (1961)

McCabe, A. M., 'Quaternary deposits and glacial stratigraphy in Ireland', *Quaternary Science Reviews* **6**, 259–99 (1987)
— Coope, R., Gennard, D. E. and Doughty, P., 'Freshwater organic deposits and stratified sediments between Early and Late Midlandian (Devensian) till sheets, at Aghnadarragh, County Antrim, Northern Ireland', *Journal of Quaternary Science* **2**, 11–33 (1987)

McCarroll, D., 'Deglaciation of the Irish Sea Basin: a critique of the glaciomarine hypothesis', *Journal of Quaternary Science* **16**, 393–404 (2001)

McCormick, F., 'Early faunal evidence for dairying', *Oxford Journal of Archaeology* **11**, 201–9 (1992)
— 'Cows, ringforts and the origins of Early Christian Ireland', *Emania* **13**, 33–37 (1995)
— 'Early evidence for wild animals in Ireland', in N. Benecke (ed.), *The Holocene History of the European Vertebrate Fauna*, 355–72 (Rahden, Verlag Marie Leidorf GmBH, 1999)
— 'The horse in early Ireland', *Anthropozoologica* **42**, 85–104 (2007)
— 'The decline of the cow; agricultural and settlement change in early medieval Ireland', *Peritia* **20**, 209–24 (2008)
— and Murray, E., *Knowth and the Zooarchaeology of Early Christian Ireland* (Dublin, Royal Irish Academy, 2007)

McCracken, E., 'Charcoal-burning ironworks in seventeenth and eighteenth century Ireland', *Ulster Journal of Archaeology* **20**, 123–38 (1955)
— 'The woodlands of Donegal 1600–1840', *Donegal Annual* **4**, 62–4 (1958)
— *The Irish Woods since Tudor Times; their Distribution and Exploitation* (Newton Abbot, David & Charles, 1971)
— and McCracken, D., *A Register of Trees for County Londonderry, 1768–1911* (Belfast, Public Records Office of Northern Ireland, 1984)

McDermott, F., Mattey, D. O. and Hawkesworth, C., 'Centennial-scale Holocene climate variability revealed by high-resolution speleothem D ^{18}O record from SW Ireland', *Science* **294**, 1328–31 (2001)

McErlean, T., McConkey, R. and Forsythe, W. (eds), *Strangford Lough; an Archaeological Survey of the Maritime Cultural Landscape*, Environment and Heritage Service, Northern Ireland Archaeological monographs 6 (Belfast, The Blackstaff Press, 2002)

MacLean, R., 'Eat your greens: an examination of the potential diet available in Ireland during the Mesolithic', *Ulster Journal of Archaeology* **56**, 1–8 (1993)

McMillan, N., 'Quaternary deposits around Lough Foyle, North Ireland', *Proceedings of the Royal Irish Academy* **58B,** 185–205 (1957)

McNally, A. and Doyle, G. T., 'A study of sub-fossil pine layers in a raised bog complex in the Irish midlands: 1. Palaeowood extent and dynamics', *Proceedings of the Royal Irish Academy* **84B**, 57–81 (1984)

McVicker, S. J. and Hall, V. A., 'Recent landscape history of Slieve Muck, Mourne Mountains, Co. Down', *Irish Naturalists' Journal* **25**, 353–8 (1997)

Mighall, T. M. and Chambers, F. M., 'The environmental impact of prehistoric mining in Copa Hill, Cwmystywyth, Wales', *The Holocene* **3**, 260–4 (1993)
— Chambers, F. M., Lanting, J. M. and O'Brien W. F., 'Prehistoric copper mining and its importance on vegetation: palaeoecological evidence from Mount Gabriel, Co. Cork, southwest Ireland', in R.A. Nicholson and T. P. O'Connor (eds), *People as an Agent of Environmental Change* (Oxford, Oxbow, 2000)
— Legeard, J. G. A., Chambers, F. M., Field, M. H. and Mahl, P., 'Mineral deficiency and the presence of *Pinus sylvestris* on mires during the mid- to late-Holocene: palaeoecological data from Cadogan's Bog, Mizen Peninsula, Co. Cork, southwest Ireland', *The Holocene* **14**, 95–109 (2004)

Mitchell, F. J. G., The vegetational history of the Killarney oak woods, SW Ireland: evidence from fine spatial resolution pollen analysis', *Journal of Ecology* **76**, 415–36 (1988)
— 'The history and vegetation dynamics of a yew wood (*Taxus baccata* L.) in SW Ireland', *New Phytologist* **115**, 573–7 (1990)
— 'The impact of grazing and human disturbance on the dynamics of a woodland in SW Ireland', *Journal of Vegetational Science* **1**, 245–54 (1990)
— 'The biogeographical implications of the distribution and history of the strawberry tree, *Arbutus unedo*, in Ireland', in M. J. Costello and K. S. Kelly (eds), *Biogeography of Ireland: Past, Present and Future*, 35–44 (Dublin, Irish Biogeographical Society, 1993)
— 'The dynamics of Irish post-glacial forests', in J. R. Pilcher and S. S. Mac an tSaoir (eds), *Wood, Trees and Forests in Ireland*, 13–22 (Dublin, Royal Irish Academy, 1995)
— 'Natural invaders: the postglacial tree colonisation of Ireland', in C. Moriarity and D. Murray (eds), *Biological Invaders; the Impact of Exotic Species*, 2–12 (Dublin, Royal Irish Academy, 2002)
— 'How open were European primeval forests? Hypothesis testing using palaeoecological data', *Journal of Ecology* **93**, 168–70 (2005)
— and Bradshaw, R. H. W., 'The recent history of native woodland in Killarney, SW Ireland', *British Ecological Society Bulletin* **15**, 18–19 (1984)

Mitchell, G. F., 'Studies in Irish Quaternary deposits. Some lacustrine deposits near Dunshaughlin, County Meath', *Proceedings of the Royal Irish Academy* **46B**, 13–37 (1940)
— 'Studies in Irish Quaternary deposits: No 2. Some lacustrine deposits near Rathoath, Co. Meath. The reindeer in Ireland', *Proceedings of the Royal Irish Academy* **46B**, 173–82 (1941)
— 'Studies in Irish Quaternary deposits: No 7', *Proceedings of the Royal Irish Academy* **53B**, 111–206 (1951)
— 'Further identification of macroscopic plant fossils from Irish Quaternary deposits, especially from a Late-glacial deposit at Mapastown, Co. Louth', *Proceedings of the Royal Irish Academy* **55B**, 225–81 (1953)
— 'Studies in Irish Quaternary deposits: No 9. A pollen diagram from Lough Gur, County Limerick', *Proceedings of the Royal Irish Academy* **56C**, 481 (1954)
— 'Studies in Irish Quaternary deposits: No. 10. The Mesolithic site at Toome Bay, Co. Londonderry', *Ulster Journal of Archaeology* **18**, 1–16 (1955)
— 'Studies in Irish Quaternary deposits: No. 11', *Proceedings of the Royal Irish Academy* **57B**, 185–250 (1956)
— 'Late-glacial finds of *Lepidurus arcticus* (Pallas) in the British Isles', *Nature* **180**, 513 (1957)
— 'Moranic ridges on the floor of the Irish Sea', *Irish Geography* **4**, 335–44 (1963)
— 'Littleton Bog, Tipperary: an Irish agricultural record', *Journal of the Royal Society of Antiquaries of Ireland* **95**, 121–3 (1965)
— 'Littleton Bog, Tipperary: an Irish vegetational record', Geological Society of America Special Paper 84, in H. E. Wright and D. G. Fry (eds), *International Studies on the Quaternary*, VIIth Congress of the International Association for Quaternary Studies, 1–16 (1965)

— *The Irish Landscape* (London, Collins, 1976)

— 'Periglacial Ireland', *Philosophical Transactions of the Royal Society of London* **B 280**, 199–209 (1977)

— 'Tillage tools in Ireland during the past 5,000 years', *Journal of Earth Sciences of the Royal Dublin Society* **2**, 27–40 (1979)

— *The Shell Guide to Reading the Irish landscape* (Dublin, Country House, 1986)

— *Man and Environment in Valencia Island* (Dublin, Royal Irish Academy, 1989)

— and Parkes, H. M., 'Studies in Irish Quaternary deposits: No. 6. The Giant Deer in Ireland', *Proceedings of the Royal Irish Academy* **52B**, 291–381 (1949)

— and Watts, W. A., 'The history of the Ericaceae in Ireland during the Quaternary epoch', in D. Walker and R. G. West (eds), *Studies in the Vegetational History of the British Isles* (Cambridge, Cambridge University Press, 1970)

— and Coxon, P., *Landscape Archaeology in South-west Valencia Island, a Field Excursion Guide* (Dublin, Trinity College, 1985)

Moen, R. A., Pastor, J. and Cohen, Y., 'Antler growth and the extinction of the Irish elk', *Evolutionary Ecology Research* **1**, 235–49 (1999)

Moffatt, C. B., 'The mammals of Ireland', *Proceedings of the Royal Irish Academy* **44B**, 61–128 (1938)

Molloy, K. and O'Connell, M., 'The nature of the vegetational changes at about 5000 BP with particular reference to the elm decline; fresh evidence from Connemara, western Ireland', *New Phytologist* **107**, 203–20 (1987)

— and O'Connell, M., 'Neolithic agriculture – fresh evidence from Cleggan, Connemara', *Archaeology Ireland* **2**, 67–70 (1988)

— and O'Connell, M., 'Palaeoecological investigations towards the reconstruction of woodland and land use history at Lough Sheeauns, Connemara, western Ireland', *Review of Palaeobotany and Palynology* **67**, 75–113 (1991)

— and O'Connell, M., 'Palaeoecological investigations towards the reconstruction of environment and land-use changes during the prehistory of Céide Fields, western Ireland', *Probleme der Kustenforschung im Sudliche Noorseegebiet* **23**, 187–225 (1995)

— and O'Connell, M., 'Holocene vegetation and land-use dynamics in the karstic environment of Inis Oirr, Aran Islands, western Ireland: pollen analytical evidence evaluated in the light of the archaeological record', *Quaternary International* **113**, 41–64 (2004)

Molyneaux, T., 'A discourse concerning the large horns frequently found under ground in Ireland, concluding from them that the great American deer call'd moose, was formerly common in that island: with remarks on some other things natural to that country', *Philosophical Transactions of the Royal Society, London*, 489–12 (1697)

Mook, W. G., 'Business meeting – Recommendations, resolutions adopted by the 12th International *Radiocarbon* Conference', *Radiocarbon* **28**, 799 (1986)

Moore, P. D., 'Tree boundaries on the move', *Nature* **326**, 545 (1987)

Moriarity C. and Murray, D. (eds), *Biological Invaders; the Impact of Exotic Species* (Dublin, Royal Irish Academy, 2002)

Morrison, M. E. S., 'The water balance of the raised bog', *Irish Naturalists' Journal* **11**, 303–8 (1955)

— 'Factors in the degeneration of the prehistoric woodland', *Irish Naturalists' Journal* **12**, 57–65 (1956)

— 'Evidence and interpretation of 'Landnam' in the north-east of Ireland', *Botaniska Notiser* **12**, 185 (1959)

— 'The palynology of Ringneill Bay, a new Mesolithic site in Co. Down, Northern Ireland', *Proceedings of the Royal Irish Academy* **61C**, 171–82 (1961)

— and Stephens, N., 'Stratigraphical and pollen analysis of the raised beach deposit at Ballyhalbert, Co. Down, Northern Ireland', *New Phytologist* **59**, 153–66 (1960)

— and Stephens, N., 'A submerged Late-Quaternary deposit at Roddans Port on the north-east coast of Ireland', *Philosophical Transactions of the Royal Society of London* **249B**, 221–55 (1965)

Muir, G., Lowe, A. J., Fleming, C. C. and Vogl, C., 'High nuclear genetic diversity, high levels of outcrossing and low differentiation among remnant populations of *Quercus petraea* at the margin of its range in Ireland', *Annals of Botany* **93**, 691–7 (2004)

Murray, E., McCormick, F. and Plunkett, G., 'The food economies of Atlantic island monasteries: the documentary and archaeo-environmental evidence', *Environmental Archaeology* **9**, 179–88 (2004)

Nelson, C., *The Cause of the Calamity; Potato Blight in Ireland and the Role of the National Botanic Gardens, Glasnevin* (Dublin, The Stationery Office, 1995)

Noe-Nygaard, N., Price, T. D. and Hede, S. U., 'Diet of aurochs and early cattle in southern Scandinavia: evidence from ^{15}N and ^{13}C stable isotopes', *Journal of Archaeological Science* **32**, 855–71 (2005)

O'Connell, M., 'The developmental history of Scragh Bog, Co. Westmeath and the vegetational history of its hinterland', *New Phytologist* **85**, 301–19 (1980)

— 'Reconstruction of local landscape development in the post-Atlantic based on palaeoecological investigations at Carrownaglogh prehistoric field system, County Mayo, Ireland', *Review of Palaeobotany and Palynology* **49**, 117–76 (1986)

— 'Early cereal-type pollen records from Connemara, Western Ireland and their possible significance', *Pollen et Spores* **19**, 207–24 (1987)

— 'Early land use in north-east County Mayo – the palaeoecological evidence', *Proceedings of the Royal Irish Academy* **90C**, 259–79 (1990)

— 'Origins of Irish lowland blanket peat', in G. J. Doyle (ed.), *Ecology and Conservation of Irish Peatlands*, 49–71 (Dublin, Royal Irish Academy, 1990)

— 'Vegetational and environmental changes in Ireland during the later Holocene', in M. O'Connell (ed.), *The Post Glacial Period (10,000–0 BP); Fresh Perspectives*, 21–6 (Dublin, Irish Association for Quaternary Studies, 1991)

— Mitchell, F. J. G., Readman, P. W., Doherty, T. J. and Murray, D. A., 'Palaeoecological investigation towards the reconstruction of the post-glacial environment at Lough Doo, County Mayo, Ireland', *Journal of Quaternary Science* **2**, 149–64 (1987)

— Molloy, K and Bowler, M., 'Post-glacial landscape evolution in Connemara, western Ireland with particular reference to woodland history', in H. H. Birks, H. J. B. Birks, P. E. Kaland and D. Moe (eds), *The Cultural Landscape: Past, Present and Future*, 267–87 (Cambridge, Cambridge University Press, 1988)

— and Ní Ghrainne, E., 'Palaeoecology [of Inishbofin]', in P. Coxon and M. O'Connell (eds), *Clare Island and Inishbofin, Field Guide* **17**, 61–103 (Dublin, Irish Association for Quaternary Studies, 1991)

— M., Huang, C. C. and Eicher, U., 'Multidisciplinary investigations, including stable-isotope studies, of thick Late-glacial sediments from Tory Hill, Co. Limerick, western Ireland', *Palaeogeography, Palaeoclimatology and Palaeoecology* **147**, 169–208 (1999)

— and Molloy, K., 'Farming and woodland dynamics in Ireland during the Neolithic', *Proceedings of the Royal Irish Academy, Biology and Environment* **101B**, 99–128 (2001)

O'Corrain, D., *Ireland before the Normans* (London, Gill & Macmillan, 1972)

O'Croinin, D., *Early Medieval Ireland 400–1200* (London, Longman, 1995)

O'Sullivan, A., 'Trees, woodland and woodmanship in Early Medieval Ireland (1994)', *Botanical Journal of Scotland* **46**, 674–81 (1994)

— *The Archaeology of Lake Settlement in Ireland*, Discovery Programme Monograph 4 (Dublin, Irish Discovery Programme and Royal Irish Academy, 1998)

O'Sullivan, P. E., Oldfield, F. and Battarbee, R. W., 'Preliminary studies of Lough Neagh sediments. 1. Stratigraphy, chronology and pollen analysis', in H. J. B. Birks and R. G. West (eds), *Quaternary Plant Ecology*, 109–15 (Oxford, Blackwell Scientific Publications, 1973)

Overland, A. and O'Connell, M., 'Fine-spatial palaeoecological investigations towards reconstructing late Holocene environmental change, landscape evolution and farming activity in Barrees, Beara Peninsula, southwestern Ireland', *Journal of the North Atlantic* **1**, 37–73 (2008)

Parker, A. G., Goudie, A. S., Anderson, D. E., Robinson, M. A. and Bonsall, C., 'A review of the mid-Holocene elm decline in the British Isles', *Progress in Physical Geography* **26**, 1–45 (2002)

Parkes, H. M. and Mitchell, F. J. G., 'Vegetation history at Clonmacnoise, Co. Offaly', *Proceedings of the Royal Irish Academy, Biology and Environment* **100B**, 35–40 (2000)

Pearson, G. W., Pilcher, J. R., Corbett, D. M. and Qua, F., 'High-precision ^{14}C measurement of Irish oaks to show the natural ^{14}C variations from D 1840 to 5210 BC', *Radiocarbon* **28**, 911–34 (1986)

Pennington, W., 'Lags in adjustment of vegetation to climate caused by the pace of soil development; evidence from Britain', *Vegetatio* **67**, 105–18 (1986)

Peterken, G. F., *Natural Woodland: Ecology and Conservation in Northern Temperate Regions* (Cambridge, Cambridge University Press, 1996)

Piggott, C. D., '*Tilia cordata* Miller', *Journal of Ecology* **79**, 1147–207 (1991)

Pilcher, J. R., 'Archaeology, palaeoecology and radiocarbon dating of the Beaghmore stone circle site', *Ulster Journal of Archaeology* **32**, 73–92 (1969)
— 'Pollen analysis and radiocarbon dating of a peat on Slieve Gallion, Co. Tyrone', *New Phytologist* **72**, 681–9 (1973)
— 'Speculations on Neolithic land clearance', *Irish Archaeological Forum* **11**, 1–6 (1975)
— Smith, A. G., Pearson, G. W. and Crowder, A., 'Land clearance in the Irish Neolithic: new evidence and interpretation', *Science* **172**, 560–2 (1971)
— and Smith, A.G., 'Palaeoecological investigations at Ballynagilly, a Neolithic and Bronze Age settlement in County Tyrone, Northern Ireland', *Philosophical Transactions of the Royal Society of London* **286B**, 345–69 (1979)
— and Larmour, R., 'Late-glacial and post-glacial vegetational history of the Meenadoan Nature Reserve, County Tyrone', *Proceedings of the Royal Irish Academy* **82B**, 278–95 (1982)
— and Hall, V. A., 'Towards a tephrochronology for the Holocene of the north of Ireland', *The Holocene* **2**, 255–9 (1992)
— Baillie, M. G. L., Brown, D. M., McCormac, F. G., MacSweeney, P. B. and Lawrence, A. S., 'Dendrochronology and subfossil pine in the north of Ireland', *Journal of Ecology* **83**, 665–71 (1995)
— and Hall, V. A., *Flora Hibernica* (Cork, The Collins Press, 2001)

Platts, E. A. and Speight, M. C. D., 'The taxonomy and distribution of the Kerry slug, *Geomalacus maculosus* Allman 1843 (Mollusca: Arionadae) with a discussion of its native status as a threatened species', *Irish Naturalists' Journal* **22**, 417–30 (1988)

Plunkett, G. M., 'Tephra-linked peat humification records from Irish ombrotrophic bogs questions nature of sear forcing at 850 cal. yr. BC', *Journal of Quaternary Science* **21**, 9–16 (2006)
— Pilcher, J. R., McCormac, F. G. and Hall, V. A., 'New dates for first millennium BC tephra isochrones in Ireland', *The Holocene* **14**, 780–6 (2004)
— Whitehouse, N. J., Hall, V. A., Charman, D. J., Blaauw, M., Kelly, E. and Mulhall, I., 'A multi-proxy palaeoenvironmental investigation of the findspot of an Iron Age bog body from Oldcroghan, Co. Offaly, Ireland', *Journal of Archaeological Sciences* **36**, 265–77 (2008)

Pococke, R., *Pococke's Tour in Ireland in 1752* (Dublin, Hodges Figgis, 1891)

von Post, L., 'The geographical survey of Irish bogs', *Irish Naturalists' Journal* **6**, 210–17 (1936)

Power, C., 'Reconstructing patterns of health and dietary change in Irish populations', *Ulster Journal of Archaeology* **56**, 9–17 (1993)

Praeger, R. L., 'Post-Tertiary beds at Ballyhalbert, Co. Down', *Irish Naturalist* **6**, 306 (1897)
— 'On the buoyancy of the seeds of some Brittanic plants', *Scientific Proceedings of the Royal Dublin Society* **14**, 13–62 (1913)
— 'Recent views bearing on the problem of the Irish flora and fauna', *Proceedings of the Royal Irish Academy* **41B**, 125–45 (1932)
— 'The committee for Quaternary research in Ireland and its work; the botanical importance of the work and the inception of the present scheme', *Irish Naturalists' Journal* **5**, 126–8 (1934)
— *The Way that I Went* (Dublin, A. Figgis, 1937).

Preece, R. C. and Robinson, J. E., 'Molluscan and ostracod faunas from postglacial tufaceous deposits in County Offaly', *Proceedings of the Royal Irish Academy* **82B**, 115–31 (1982)
— Coxon, P. and Robinson, J. E., 'New biostratigraphic evidence of the postglacial colonisation of Ireland and for Mesolithic forest disturbance', *Journal of Biogeography* **13**, 487–509 (1986)

Preston, C. D., Pearman, D. A. and Hall, A. R., 'Archaeotypes in Britain', *Botanical Journal of the Linnean Society* **145**, 257–94 (2004)

Quirke, B., *Killarney National Park, a Place to Treasure* (Cork, The Collins Press, 2001)

Radcliff, T., *Report of the Agriculture and Livestock of the County of Kerry* (Dublin, W. Porter, 1814)

Raftery, B., *Trackways Through Time. Archaeological Investigations on Irish Bog Roads, 1985–1989* (Dublin, Headline Publishing, 1990)

Reader, J., *Propitious Esculent; the Potato in World History*, 1–315 (London, Heinemann, 2008)

Reid, C., 'The relation of the present plant population of the British Isles to the glacial period', *The Irish*

Naturalist **20**, 201–9 (1911)

Richards, M. P., Schulting, R. J. and Hedges, R. E. M., 'Sharp shift in diet at onset of Neolithic', *Nature* **425**, 366 (2003)

Robinson, M., 'Coleopteran evidence for the elm decline, Neolithic activity in woodland, clearance and the use of the landscape', in A. S. Fairbairn (ed.), *Plants in Neolithic Britain and Beyond*, Neolithic Studies Group Seminar Papers **5**, 27–35 (Oxford, Oxbow Books, 2000)

Robinson, P. S., 'The spread of hedged enclosure in Ulster', *Ulster Folklife* **23**, 57–69 (1977)

Roche, J. R., Mitchell, F. J. G. and Waldren, S., 'Plant community ecology of *Pinus sylvestris*, an extirpated species reintroduced to Ireland', *Biological Conservation* **18**, 2185–203 (2009)

Rohan, P. K., *The Climate of Ireland* (Dublin, Meteorological Service, 1975)

Rohling, E. J. and Palike, H., 'Centennial-scale climate cooling with a sudden cold event around 8,200 years ago', *Nature* **434**, 975–9 (2005)

Ryan, K., 'Holes and flaws in medieval Irish manuscripts', *Peritia* **6–7**, 243–64 (1987–88)

Ryan, M., 'Lough Boora excavations', *An Talaca Journal* **2**, 13–14 (1978)
— 'Archaeological excavation at Lough Boora, Boughal Townland, Co. Offaly, 1977', *Proceedings of the 7th International Peat Congress*, Dublin 407–11 (1984)

Rymer, L., 'The use of uniformitarianism and analogy in palaeoecology, particularly in pollen diagrams', in D. Walker and J. C. Guppy (eds), *Biology and Quaternary environments*, 245–57 (Canberra, Australian Academy of Science, 1978)

Sarnthein, M., Winn, K., Jung, S. J. A., Duplessy, J-C., Labeyrie, L., Erlenkauser, H. and Ganssen, G., 'Changes in east Atlantic deepwater circulation over the last 30,000 yrs: eight time slice reconstructions', *Palaeoceanography* **9**, 209–67 (1994)

Savage, R. J. G., 'Irish Pleistocene mammals', *Irish Naturalists' Journal* **15**, 117–30 (1966)

Schulting, R. J. and Richards, M. P., 'Finding the coastal Mesolithic in southwest Britain: AMS dates and stable isotope results on human remains from Caldey Island, south Wales', *Antiquity* **76**, 1001–25 (2002)

Selby, K. A., O'Brien, C. E., Brown, A. G. and Stuitjs, I., 'A multi-proxy study of Holocene lake development, lake settlement and vegetation history on central Ireland', *Journal of Quaternary Science* **20**, 147–68 (2005)

Shaw, J. and Carter, R. W. G., 'Coastal peats from northwest Ireland: implications for late-Holocene relative sea-level change and coastal evolution', *Boreas* **23**, 74–91 (1994)

Singh, G., 'Pollen analysis of a deposit at Roddans Port, Co. Down, N. Ireland, bearing reindeer antler fragments', *Grana Palynologica* **4**, 466–74 (1963)
— 'Late-glacial vegetational history of Lecale, Co. Down', *Proceedings of the Royal Irish Academy* **69B**, 189–216 (1970)
— and Smith, A. G., The post-glacial marine transgression in N. Ireland. Conclusions from estuarine and raised beach deposits: a contrast', *Palaeobotanist* **15**, 230–4 (1966)
— and Smith, A. G., 'Post-glacial vegetational history and relative land- and sea-level changes in Lecale, Co. Down', *Proceedings of the Royal Irish Academy* **73B**, 1–56 (1973)

Sleeman, D. P., 'Rabies – past, present and future', in C.V. Holland (ed.), *Modern Perspectives on Zoonoses*, 95–105 (Dublin, Royal Irish Academy, 1997)
— Devoy, R. J. and Woodman, P. C., *Proceedings of the Postglacial Colonization Conference*, Occasional publication of the Irish Biogeographical Society (1986)

Smith, A. G., 'Pollen analytical investigations of the mire at Fallahogy Td., Co. Derry', *Proceedings of the Royal Irish Academy* **59B**, 329–43 (1958)
— 'The Atlantic–Sub-Boreal transition', *Proceedings of the Linnean Society of London* **172**, 38–49 (1961)
— 'Cannons Lough, Kilrea, County Derry: stratigraphy and pollen analysis', *Proceedings of the Royal Irish Academy* **61B**, 369–83 (1961)
— 'Problems in the study of the earliest agriculture in Northern Ireland', *Report of the VIth International Congress on the Quaternary*, Warsaw 1961, Palaeobotanic Section **2**, 461–71 (1964)
— 'Problems of threshold and inertia related to post-glacial habitat change', *Proceedings of the Royal Society* **161B**, 331–42 (1965)
— 'The influence of Mesolithic and Neolithic man on British vegetation: a discussion', in D. Walker and

R. G. West (eds), *Studies in the Vegetational History of the British Isles*, 81–96 (Cambridge, Cambridge University Press, 1970)

— 'Late-glacial and post-glacial vegetational and climatic history of Ireland: a review', in N. Stephens and R. E. Glasscock (eds), *Irish Geographical Studies*, in honour of E. Estyn Evans, 65–88 (Belfast, Department of Geography, Queen's University, 1970)

— 'Neolithic and Bronze Age landscape change in northern Ireland', in J. G. Evans, S. Limbrey and H. Cleeve (eds), *The Effect of Man on the Landscape: the Highland Zone*, Council for British Archaelogy, Report **11**, 64–74 (1975)

— 'Palynology of a Mesolithic-Neolithic site in Co. Antrim', *IV International Palynolological Conference*, Lucknow (1976–77) **3**, 248–57 (1981)

— 'Newferry and the Boreal-Atlantic transition', *New Phytologist* **38**, 35–55 (1984)

— and Willis, E. H., 'Radiocarbon dating of the Fallahogy landnam phase', *Ulster Journal of Archaeology* **24–25**, 16–24 (1962)

— Pilcher, J. R. and Pearson, G. W., 'New radiocarbon dates from Ireland', *Antiquity* **45**, 97–100 (1971)

— and Collins, A. E. P., 'The stratigraphy, palynology and archaeology of diatomite deposits at Newferry, Co. Antrim, Northern Ireland', *Ulster Journal of Archaeology* **34**, 3–25 (1971)

— and Pilcher, J. R., 'Pollen analysis and radiocarbon dating of deposits at Slieve Gullion passage grave, Co. Armagh', *Ulster Journal of Archaeology* **35**, 17–21 (1972)

— and Pilcher, J. R., 'Radiocarbon dates and the vegetational history of the British Isles', *New Phytologist* **72**, 903–14 (1973)

— Gaskell Brown, C., Goddard, I. C., Pearson, G. W. and Dresser, P. Q., 'Archaeology and environmental history of a barrow at Pubble, Loughermore townland, County Londonderry', *Proceedings of the Royal Irish Academy* **81C**, 2–66 (1981)

— and Goddard, I. C., 'A 12,500 year record of vegetational history at Sluggan Bog, Co. Antrim, N. Ireland (incorporating a pollen zone scheme for the non-specialist)', *New Phytologist* **118**, 167–97 (1991)

Speight, M. C. D., 'The extinction of indigenous *Pinus sylvestris* in Ireland: relevant faunal data', *Irish Naturalists' Journal* **21**, 449–53 (1985)

Speller, G., 'Spatial variability of molluscan successions from the early Holocene of Ireland', *Quaternary Newsletter* **105**, 61–2 (2005)

Stelfox, A.W., 'The problem of the Irish elk', *Irish Naturalists' Journal* **5**, 74–6 (1934)

— Kuiper, J. G. J., McMillan, N. and Mitchell, G. F., 'The late-glacial and post-glacial Mollusca of the White Bog, Co. Down', *Proceedings of the Royal Irish Academy* **72B**, 185–207 (1972)

Stephens, N., 'Some observations on the "interglacial" platform and the early post-glacial raised beach on the east coast of Ireland', *Proceedings of the Royal Irish Academy* **58B**, 129–49 (1957)

— and Synge, F. M., 'A Quaternary succession at Sutton, Co. Dublin', *Proceedings of the Royal Irish Academy* **59B**, 19–27 (1958)

— and Collins, A. E. P., 'The Quaternary deposits at Ringneill Quay and Ardmillan, Co. Down', *Proceedings of the Royal Irish Academy* **61C**, 41 (1960)

Stevenson, A. C. and Thompson, D. B. A., 'Long-term changes in heather moorland in upland Britain and Ireland: palaeoecological evidence for its importance', *The Holocene* **3**, 70–76 (1993)

Stillman, C. J., 'The post-glacial change in sea level in south west Ireland: new evidence from fresh water deposits on the floor of Bantry Bay', *Scientific Proceedings of the Royal Dublin Society* **A3**, 125–7 (1968)

Stout, M., *Early Christian Ireland, Settlement and Environment, a History of Settlement in Ireland* (Dublin, Trinity College, 1997)

Stout, G. and Stout, M., 'Early landscapes: from prehistory to plantation', in F. H. A. Aalen, K. Whelan and M. Stout (eds), *Atlas of the Rural Irish Landscape*, 31–63 (Cork, Cork University Press and University of Toronto Press, 1997)

Stuart, A. J., 'Insularity and Quaternary vertebrate faunas in Britain and Ireland', in R. C. Preece (ed.), *Island Britain: a Quaternary Perspective*, 111–25 (London, The Geological Society, 1995)

— Kosintsev, P. A., Higham, T. F. G. and Lister, A. M., 'Pleistocene to Holocene extinction dynamics in giant deer and woolly mammoth', *Nature* **431**, 684–9 (2004)

Synge, F. M., 'Records of sea levels during the Late Devensian', *Philosophical Transactions of the Royal Society of*

London **B 280**, 211–28 (1978)

— and Stephens, N., 'The Quaternary period in Ireland – an assessment', *Irish Geography* **4**, 121–30 (1960)

Synnott, C. and Downey, L., 'Bog butter; its historical context and chemical composition', *Archaeology Ireland* **18**, 32–5 (2004)

Tallantire, P. A., 'The early-Holocene spread of hazel (*Corylus avellana* L.) in Europe north and west of the Alps: an ecological hypothesis', *The Holocene* **12**, 81–96 (2002)

Taylor, J. A. 'The peatlands of Great Britain and Ireland', in A. J. P. Gore (ed.), *Ecosystems of the World 4B. Mires; Swamp, Bog, Fen and Moor*, 1–46 (Amsterdam, Elsevier Scientific Publications, 1983)

Teunissen, D. and Teunissen-van Oorschot, H. G. C. M., 'The history of the vegetation in SW Connemara (Ireland)', *Acta Botanica Neerlandica* **29**, 285–306 (1980)

Thompson, P. and Maloney, B. K., 'The palaeoenvironments of Coolarken Pollnagollum (Pollnagollum of the Boats) Cave, County Fermanagh, Northern Ireland: evidence from phytolith analysis', *Cave Science* **20**, 13–15 (1993)

Tomlinson, R. W., 'The erosion of peat in the uplands of Northern Ireland', *Irish Geography* **14**, 51–64 (1981)

Turney, C. S. M., Coope, G. R., Harkness, D. D., Lowe, J. J. and Walker, M. J. C., 'Implications for the dating of Wisconsian (Weichselian) late-glacial events of systematic radiocarbon age differences between terrestrial plant macrofossils from a site in SW Ireland', *Quaternary Research* **53**, 114–21 (2000)

Von Engelbrechten, S., McGee, S., Little, D. J. and Mitchell, F. J. G., 'A palaeoecological study of Blackloon Wood, Co. Mayo. Vegetation dynamics and human impact through the Holocene period (c. 10,000 years to present)', *Forest Systems Research Group Report* **42** (2000)

Vuorela, I., 'The indication of farming in pollen diagrams from southern Finland', *Acta Botanica Fennica* **87**, 3–40 (1980)

Waddel, J., *The Prehistoric Archaeology of Ireland* (Galway, Galway University Press, 1998)

Walker, M. J. C., Coop, G. R. and Lowe, J. J., 'The Devensian (Weichselian) Late-glacial in northwest Europe (Ireland, Britain, north Belgium, The Netherlands, northwest Germany)', *Journal of Quaternary Science* **9**, 109–18 (1993)

— Johnsen, S., Rasmussen, S. O., Popp, T., Steffensen, J-P., Gibbard, P., Hoek, W., Kershaw, P., Kroner, B., Litt, T., Lowe, D. J., Nakagawa, T., Newnham, R. and Schwander, J., 'Formal definition and dating of the GSSP (Global Stratotype Section and Point) for the base of the Holocene using the Greenland NGRIP ice core and selected auxiliary records', *Journal of Quaternary Science* **24**, 3–17 (2009)

Warren, G., 'Life in the trees; Mesolithic people and the woods of Ireland', *Archaeology Ireland* **17**, 20–3 (2003)

Watts, W. A., 'Post-Atlantic forests in Ireland', *Proceedings of the Linnean Society* **172**, 33 (1961)

— 'Late-glacial pollen zones in western Ireland', *Irish Geography* **4**, 367–75 (1963)

— 'The late Devensian vegetation of Ireland', *Philosophical Transactions of the Royal Society of London* **B 280**, 273–93 (1977)

— 'Contemporary accounts of the Killarney Woods 1580–1870', *Irish Geography* **17**, 1–13 (1984)

— 'The Holocene history of the Burren, western Ireland', in E. Y. Haworth and J. W. G. Lund (eds), *Lake Sediments and Environmental History*, 359–76 (Leicester, Leicester University Press, 1984)

— 'Quaternary vegetation cycles', in K. J. Edwards and W. P. Warren (eds), *The Quaternary History of Ireland*, 155–185 (London, Academic Press, 1985)

Webb, D. A., 'The biological vice-counties of Ireland', *Proceedings of the Royal Irish Academy* **80B**, 179–96 (1980)

— 'The flora of Ireland in its European context', *Journal of Life Sciences, Royal Dublin Society* **4**, 143–60 (1983)

— Parnell, J. and Doogue, D., *An Irish Flora* (Dundalk, Dundelgan Press, 1996)

Weinstock, J., 'Geographical variation of reindeer (*Rangifer tarandus*) in Europe during the Late Glacial', in N. Benecke (ed.), *The Holocene History of the European Vertebrate Fauna*, 283–94 (Rahden, VML Verlag Marie Leidorf GmbH, 1999)

Weir, D. A., 'Dark Ages and the pollen record', *Emania* **11**, 21–30 (1993)

— 'A palynological study of landscape development in County Louth from the second millennium BC to the first millennium AD', in *Discovery Programme Reports* **2**, *Project Results 1993*, 77–126 (Dublin, Royal Irish Academy, 1995)

Whelan, C. B., 'Pollen analysis and Irish prehistory', *Irish Naturalists' Journal* **5**, 134–7 (1934)

Whitehouse, N. J. and Smith, D. N., '"Islands" in Holocene forests: implications for forest openness, landscape clearance and "culture-steppe" species', *Environmental Archaeology* **9**, 199–208 (2004)
— 'The Holocene British and Irish ancient forest fossil beetle fauna: implications for forest history, biodiversity and faunal colonisation', *Quaternary Science Reviews* **25**, 1755–89 (2006)
— Murphy, E. and Plunkett, G., 'Human exploitation and the biota of islands', *Environmental Archaeology* **9**, 113–15 (2004)

Wickham-Jones, C. R. and Woodman, P. C., 'Studies on the early settlement of Scotland and Ireland', *Quaternary International* **49/50**, 13–20 (1998)

Williams, E., 'Dating the introduction of food production into Britain and Ireland', *Antiquity* **63**, 510–21 (1989)

Wilson, P., 'The postglacial colonization of Ireland by fish, amphibians and reptiles', in P. Sleeman, R. J. N. Devoy and P. C. Woodman (eds), *The Proceedings of the Postglacial Colonization Conference*, Occasional publications of the Irish Biographical Society No. **1.**, 53–8 (1984)
— and Carter, R. W. G., *Northwest Co. Donegal and Northwest Co. Londonderry*, Field Guide **7**, 56 (Dublin, Irish Association for Quaternary Studies, 1984)
— and Bradley, S. M., 'Development and age structure of Holocene coastal sand dunes at Horn Head, near Dunfanaghy, Co. Donegal, Ireland', *The Holocene* **7**, 187–98 (1997)
— McGourty, J. and Bateman, M. D., 'Mid- to late-Holocene coastal dune event stratigraphy for the north coast of Northern Ireland', *The Holocene* **14**, 406–16 (2004)

Wingfield, R. T. R., 'A model of sea-levels in the Irish and Celtic seas during end-Pleistocene to Holocene transition', in R. C. Preece (ed.), *Island Britain: a Quaternary Perspective*, 209–42 (London, The Geological Society, 1995)

Wintle, A. G., Clarke, M. L., Musson, F. M., Orford, J. D. and Devoy, R. J. N., 'Luminescence dating of recent dunes on Inch Spit, Dingle bay, southwest Ireland', *The Holocene* **8**, 331–40 (1998)

Woodman, P. C., 'Settlement patterns of the Irish Mesolithic', *Ulster Journal of Archaeology* **36 & 37**, 1–16 (1973–74)
— 'Recent excavations at Newferry, Co. Antrim', *Proceedings of the Prehistoric Society* **43**, 155–99 (1977)
— *The Mesolithic in Ireland: hunter-gatherers in an insular environment*, British Archaeological Reports, British Series **58** (Oxford, 1978)
— 'The chronology and economy of the Irish Mesolithic', in P. Mellars (ed.), *The Early Postglacial Settlement of Northern Europe*, 333–69 (London, Duckworth, 1978)

Woodman, P. C., 'A Mesolithic camp in Ireland', *Science* **245**, 120–32 (1982)
— 'Problems in the colonisation of Ireland', *Ulster Journal of Archaeology* **49**, 7–17 (1986)
— McCarthy, M. and Monaghan, N., 'The Irish Quaternary fauna project', *Quaternary Science Reviews* **16**, 129–59 (1997)
— Anderson, E. and Finlay, N. (eds), *Excavations at Ferriter's Cove*, 1983–95 (Bray, Wordwell, 1999)

Yalden, D., *The History of British Mammals* (London, Poyser, 1999)
— and Carthy, R. I., 'The archaeological record of birds in Britain and Ireland compared: extinctions or failures to arrive?', *Environmental Archaeology* **9**, 123–6 (2004)

Young, A., *A Tour in Ireland* (London, Cadell & Dodsley, 1780)

INDEX